Practical
CONSERVATION
BOUNDARY HABITATS

BY SUSAN CARR AND MARY BELL

THE OPEN UNIVERSITY IN ASSOCIATION WITH THE
NATURE CONSERVANCY COUNCIL

Hodder & Stoughton

A MEMBER OF THE HODDER HEADLINE GROUP

CONSERVATION

Open University course team

Andrew Lane (Course Team Chair)

Joyce Tait (Senior Lecturer)

Susan Carr (Lecturer)

Mary Bell (Staff Tutor)

Graham Turner (BBC Producer)

Jennie Moffatt (Course Manager)

Julie Bennett (Editor)

Lesley Passey (Designer)

Keith Howard (Graphic Artist)

Roy Lawrance (Graphic Artist)

Ray Munns (Cartographer)

Sue Snelling (Secretary)

Pat Shah (Secretary)

ISBN 0 340 53367 6

First published 1991
Impression number 12 11 10 9 8 7 6 5 4 3
Year 1999 1998 1997 1996 1995 1994

Typeset by The Open University.
Printed in Great Britain for Hodder & Stoughton Educational,
a division of Hodder Headline Plc, 338 Euston Road, London NW1 3BH
by Butler & Tanner Ltd, Frome and London

Contents

Foreword

This book is produced by the Open University as part of the *Practical Conservation* training programme which deals with all aspects of conservation on land that is managed largely for commercial or recreational purposes (see Figure 0.1).

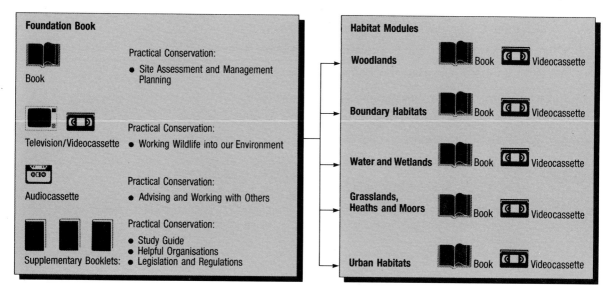

Figure 0.1
The Open University teaching programme for Practical Conservation

The foundation module covers site assessment and land use management planning in general and includes:

▶ a foundation book;

▶ a video cassette of a 50 minute television programme;

▶ a 60 minute audio cassette;

▶ two supplementary booklets;

▶ a *Study Guide* to the full programme.

This book with its accompanying 30 minute video cassette forms one of a series of modules on practical aspects of conservation management for a range of habitats:

▶ Woodlands;

▶ Boundary Habitats;

▶ Water and Wetlands;

▶ Grasslands, Heaths and Moors;

▶ Urban Habitats.

These training materials are suitable for use by groups or by individuals, studying alone or in association with a formal course. For those who would like to gain practical experience or a qualification, the Open University training programme is being incorporated into courses offered by colleges, field centres and other training bodies.

For further information please write to: Learning Materials Service Office, The Open University, PO Box 188, Walton Hall, Milton Keynes MK7 6DH.

Chapter 1
INTRODUCTION

Boundary habitats are habitats in areas that traditionally separate different land uses and different land ownership. Many incorporate a physical barrier designed to control the movement of stock. They include the hedges, stone walls, banks and ditches that surround fields, as well as the footpaths, bridleways, roadside verges, railway embankments and canal sides associated with transport routes.

Well-managed boundaries can harbour a rich variety of wildlife. Often they combine elements of woodland, scrub, grassland, aquatic and cultivated habitats, which all add to their wildlife interest. They occur throughout the countryside (as well as permeating urban areas), providing links between larger areas of habitat, such as woods. Local variations in style – the laid thorn hedges found in the Midlands, the stone walls with inset field barns in the Yorkshire Dales, the steep banks which line many Devonshire lanes – play a major part in defining the distinctive landscape character of different regions and contribute greatly to the attraction of the British countryside.

Some idea of the total length of linear boundary features is given in Figure 1.1. Except where the management is mainly under the control of a single

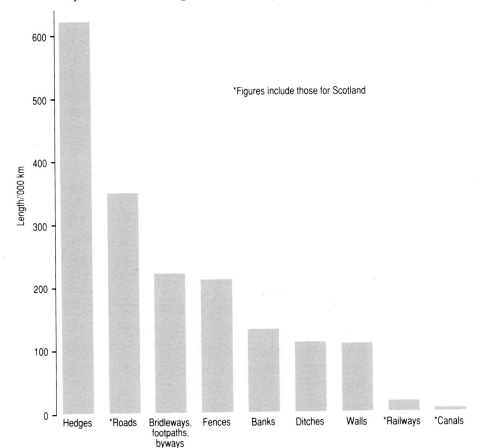

*Figures include those for Scotland

Figure 1.1 Total length of linear boundary features in England and Wales. (Sources: Figures were derived as follows. Hedges, fences, banks, ditches and walls – Countryside Commission, 1990; roads and railways – Department of Transport, Scottish Development Department and Welsh Office, 1990; bridleways, footpaths, byways – The Ramblers' Association, personal communication, 1991; canals – British Waterways, personal communication, 1991)

organisation, as in the case of roads, railways and canals, the information available is based on estimates and does not include data for Scotland. In spite of this, it is sufficient to show that boundaries represent a considerable conservation resource, providing refuges for wildlife throughout the countryside, even in otherwise intensively managed land.

The narrow, linear nature of the majority of boundary habitats can, however, make them particularly sensitive to influences from surrounding areas. In recent years, more intensive land use, with widespread use of herbicides, insecticides and fertilisers, and the practice of ploughing right up to the edge of fields, has reduced the conservation value of many boundaries on farmland. Increasing specialisation, particularly the change from mixed to all-arable farming, the use of larger machines and a steep decline in the number of people employed on the land has often meant that boundary features have been removed or their management much reduced. Pressures to reduce costs and shed labour have similarly affected the time devoted to the management of vegetation along transport routes.

 The creation of new boundary features has to some extent counterbalanced the loss of boundary habitats. The wide verges of most modern motorways provide large areas of undisturbed land, which can have considerable value for wildlife, given appropriate management. Government policies now place increasing emphasis on the environmental aspects of agriculture, encouraging renewed interest in the maintenance of existing boundary habitats on farmland, as well as in the creation of new features. In many cases, relatively small adjustments to commercial operations can result in large benefits for conservation, and restoration of the wildlife and landscape interest of some of the more degraded boundaries can be amongst the most rewarding of conservation projects to undertake.

This book is a practical guide to the conservation of boundary habitats within the framework of commercial land management. It describes how you can maintain and enhance the conservation value of existing boundary features for which you have responsibility and, where appropriate, create new ones. It shows how any potential conflicts between the functional value of boundary features and their landscape and wildlife value can be overcome by working within a management plan that takes into account both the conservation and the commercial aspects of land use. Where sympathetic management benefits wildlife and the landscape, further benefits in the form of improved public relations and enhancement of the commercial value of the land are likely to follow.

1.1   How to use this book

Like other books in the *Practical Conservation* series, this book emphasises the importance of preparing a management plan before you undertake any conservation work. This will help to clarify your ideas and ensure that others who may be involved, either currently or at some time in the future, understand fully what needs to be done. Increasingly, it is a precondition of the award of a grant.

The process of management planning is dealt with in detail in the foundation book, *Site Assessment and Management Planning*. Management planning involves first assessing the value of what you already have, identifying your objectives and constraints, exploring the management options available, choosing and setting priorities among the options and then drafting a formal plan. A summary of the stages in developing a plan and putting it into action is shown in Figure 1.2. The process is represented as cyclical, since in most cases there is no stage at which management can be said to be finished, although some operations (such as restoring a stone wall) may be longer lasting in their effects than others.

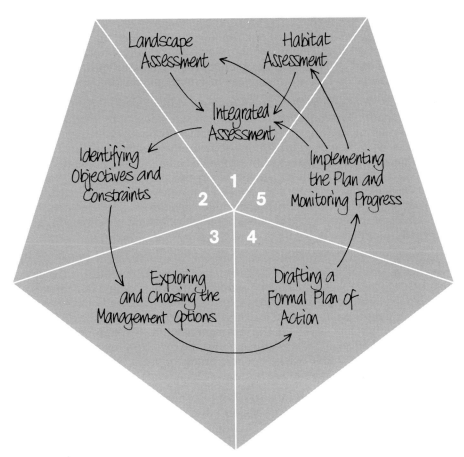

Figure 1.2   *The process of land use management planning*

This book concentrates on three of the stages in land use management planning: site assessment, choice of options and implementation. Chapters 2, 3 and 4 describe the assessment of boundary habitats for their landscape and wildlife value, and the integration of these aspects with a commercial assessment. Chapters 5 and 6 discuss some of the options available for the creation and management of boundary habitats, and Chapter 7 covers the implementation and monitoring of the chosen options. Farmland boundaries are discussed in greater detail than those along transport routes, but in many cases the same principles apply to both. Although ditches, streams and rivers could all be classed as boundary habitats, only ditches are covered here; the management of streams and rivers is described in another book in the *Practical Conservation* series, *Water and Wetlands*.

The way in which you use this book should be guided by your purpose in reading it. If you are new to conservation management or would like to improve the conservation value of your boundaries but are uncertain where to begin, it is probably best to work through the book from beginning to end. In this case, you may also find it helpful to read the foundation book. If, on the other hand, you have a specific purpose in mind, for example to revitalise a neglected hedge, or to increase the number of gamebirds nesting in the field margins, you may prefer just to select the appropriate sections of the book as you need them. The headings within the chapters and in the margins of the text are designed to help you to do this.

Throughout the book, there are practical exercises (printed on a green background) which will allow you to try out for yourself the points being made. At the end of most chapters, there is also a case study example to give further guidance on how to convert the information provided into practice.

To help with your study, there is a 30 minute video cassette which complements the book. This looks in turn at hedgerows, stone walls, ditches, verges, banks and cuttings, and field margins, examining their structure, their value as habitats and identifying some of the plants and animals that live in them. It demonstrates techniques used in boundary habitat management, such as hedge trimming, laying and coppicing, and provides examples of the ways in which people manage their boundary habitats. It concludes with some ideas for management for you to consider. Small video cassette symbols appear in the margins of the book wherever a topic is illustrated on the video cassette.

This book concentrates on the *practical* aspects of conservation. Other books, which can provide more detailed information on particular subjects, such as plant identification and the history of field margins, are listed in Appendix I. A summary of the meaning of special terms associated with boundary habitats is provided in Appendix II, and these terms are highlighted in bold type where they first appear in the text from here on. The common names for plants and animals are used throughout the main text, except in cases where no common name exists or where precise identification has practical importance; scientific names are listed in Appendix III.

Drawing up a conservation management plan need not be complicated; a simple, annotated map will do. The main thing is to be clear about what you are doing and why. Management decisions will be more soundly based if you keep in mind the following guidelines.

▶ Assess the area's existing conservation and commercial value before making changes.

▶ Identify clearly your objectives and constraints.

▶ Consider as many options as possible.

▶ Draw up a formal plan, however brief.

▶ Monitor changes in the area to check that you are achieving your objectives and be prepared to adapt your plan if necessary, but be patient: do not expect things to happen overnight.

1.2 Boundary habitat structure and terms

Regional diversity in the structure of boundary habitats is equalled or exceeded by the variety of terms used to describe them, so it is as well to start with a few definitions. This can most simply be done by first considering a stylised boundary habitat with a large number of boundary features, as shown in Figure 1.3, and then looking at some of the variations that occur.

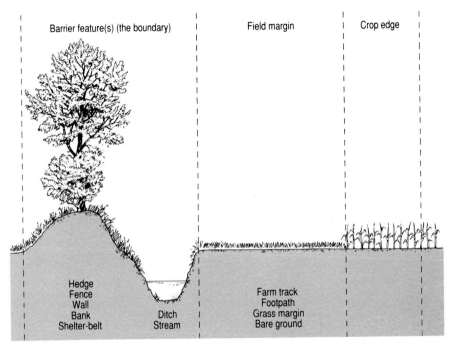

Figure 1.3   *Section across a stylised arable field boundary*

Field boundaries

Boundaries on farmland may have fields on both sides, or border a field on only one side where they lie next to a road, **green lane**, canal side or railway track. The arable field boundary shown in Figure 1.3 has three principal components:

(i) the barrier feature (in this case several features), more commonly referred to simply as 'the boundary';

(ii) the field margin separating the barrier from the crop;

(iii) the crop edge.

The barrier in this example consists of a row of shrubs and hedgerow trees planted on a bank, with a ditch alongside. In some situations, the place of the hedge may be taken by a more substantial band of trees and shrubs, forming a **shelter-belt**. In districts where stone lies on or close to the surface, the barrier may be a stone wall. In other cases, it may be simply a fence, a grass bank or a ditch. The ditch may contain water in the winter only or all year round, and the water may be stagnant or flowing. Sometimes a stream or river forms the barrier feature.

The field margin may be wide, incorporating a farm track or footpath, or almost non-existent where the crop is drilled right up to the edge of the field. It may be grassed over, or kept bare of vegetation by cultivation or herbicide treatment. The crop edge is shown as part of the field boundary in Figure 1.3 because it is often managed slightly differently from the rest of the crop and can contribute to the conservation value of the boundary. Where the field margin is narrow, the crop edge doubles as the **headland**, the area needed for turning machinery round at the ends of a field. Seed rates, pesticide use and yields in the crop edge may differ from those in the rest of the crop, because of soil compaction and other edge effects.

Field boundaries on grassland are likely to have a similar structure to that shown, except that, because they need to be stockproof, the barrier may include a fence as well as a hedge or wall, and the distinction between the field margin and the crop edge will be less clear cut.

Local terms used to describe field boundaries further enrich the variation that occurs (Figure 1.4). A ditch in Lancashire is a dyke in Norfolk and a **rhyne** or **rhine** in Somerset, while in parts of Wales the bank is known as a ditch. A **dike** or **dyke** in Scotland is similar to a stone wall in the Yorkshire Dales, and a hedge in Devon and Cornwall is a stone-faced earth bank, which may or may not be topped with a line of shrubs. It is important to be aware of such local differences when giving or acting on advice.

Scottish dyke

Cornish hedge

Somerset rhyne

Figure 1.4   Examples of regional variation in field boundaries and the terms used to describe them

Barriers such as hedges are often also a prominent feature of roadside boundaries. In addition, there is usually a verge of varying width, often containing mostly grassland species. The verge usually has a distinct outer zone, which can extend to a metre or more in width along major roads. Here, the vegetation is strongly influenced by chemical pollution resulting from car exhaust fumes and the salting of roads in winter, and by physical disturbance from traffic. The inner zone, away from the road, generally provides a more stable environment for wildlife, with taller and more diverse vegetation. Since roads need to be well drained, the verge may incorporate an open ditch.

On railway lines, the central area or **permanent way** usually has a foundation of ballast on which the sleepers and rails are laid (Figure 1.5). The permanent way, including the track and inspection walkway (the **cess path**), is normally kept free of vegetation by a weed-killing spray train. Alongside the permanent way, there is usually a verge of varying width, which may be an embankment or cutting, separated from the adjoining land by a boundary fence. Up to 5 metres at the trackside edge of the verge (the **flail strip**) is kept free of scrub and overhanging vegetation, which might otherwise obscure signals, and a strip next to the fence (the boundary strip) may be cleared periodically to allow for fence inspection and replacement.

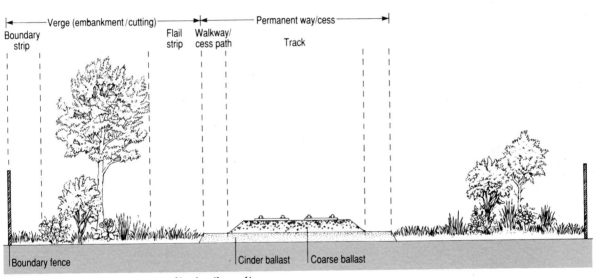

Figure 1.5   Section across a stylised railway line

The typical structure of a canal is shown in cross-section in Figure 1.6. Vegetation on the **towpath** verge is composed of a mixture of coarse grasses, tall herbs and, in places, self-sown shrubs and trees. The towpath itself is usually paved or surfaced with aggregate, and trampling further restricts plant growth on the path. However, grasses that can withstand trampling, and tall herbs typical of disturbed ground, grow on the verge between the path and the canal. Vegetation on the canal wall above the water level can be luxuriant, and reed beds occur along the submerged ledge if boat traffic is not too heavy. On the undisturbed off-side verge, ferns often flourish and the thick vegetation provides nesting sites for water fowl.

11

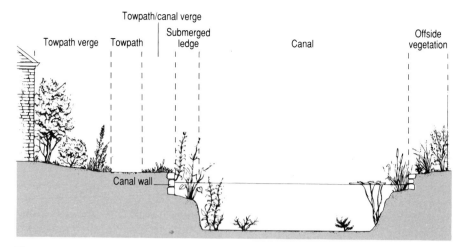

Figure 1.6   Section across a stylised canal

1.3   History of boundary features

As well as being familiar with the terms used to describe boundary features, before tackling their management you need to know something about their history, since the older a boundary is, the more valuable it is likely to be in conservation terms. Older boundary features often have a greater number of plant and animal species associated with them, make an especially important contribution to the regional character of an area and provide a vivid record of many previous generations of land management.

Field boundaries

Examples of the tremendous variation in the age of existing field patterns and boundaries are shown in Figure 1.7. In some places, evidence of very ancient field boundaries can still be seen, even though the nature of the boundary itself may have changed. Among the oldest field systems still in use are some dating from the **Bronze Age**. Examples can be seen in the Penwith peninsula of Cornwall at Zennor, where the fields are bounded with granite walls, and near Eye in Suffolk, where they are hedged. Evidence that boundaries date from at least this period is derived from their association with prehistoric sites, and occasional finds of Bronze Age hoards in the boundary banks, as well as from the very small size of the fields.

Other examples of ancient field boundaries include those in the Dengie peninsula of Essex and in the Tivetshall and Scole areas of south Norfolk. Although the precise age of these regular grids of small, rectangular fields is uncertain, they are known to date from at least the early period of **Roman settlement,** and were possibly first established much earlier than that.

Equally ancient field boundaries, although no longer in use, can still be distinguished in places as uneven marks on the ground, for example on the chalk downs of Sussex and Wiltshire and on the moors near Malham in Yorkshire. Here the position of former fields is marked by **lynchets**, large banks of soil which accumulated at the edge of the plots as a result of repeated cultivation and soil erosion on sloping land.

12

Ancient field boundaries at Zennor, Cornwall

Pre-Norman and medieval lynchets at the site of an Iron Age settlement at Malham, Yorkshire

Irregularly shaped fields typical of medieval woodland clearance, Tedburn St Mary, Devon

Curved field boundaries indicating early enclosure of open field strips, Chelmorton, Derbyshire

Ridges and furrows of medieval open fields, and subsequent enclosure, Padbury, Buckinghamshire

Post Second World War field enlargement, former hedge lines revealed by crop marks, Apsey Green Suffolk

Figure 1.7 Examples of the variation in the age of existing field boundaries

Although only scattered remnants of such ancient field systems remain visible, archaeological evidence suggests that much of the countryside was settled and farmed in established agricultural units well before the Romans arrived. It is quite likely that this existing framework was only gradually modified over successive generations, so forming the basis of many of the boundaries defined by the **Anglo-Saxon charters**, and in places still influencing the field pattern to the present day.

From time to time and in some areas, however, there has been more radical field reorganisation. Around the ninth century, the system of **open field agriculture** was introduced, in which adjacent holdings of arable land were amalgamated to create two or more large, open fields near the main settlement. Each field was farmed communally but cultivated in strips, with individual farmers renting or owning strips dispersed around the fields. Stock were grazed on common land, waste ground and on the arable fields when they were being fallowed. Such large-scale reorganisation of field boundaries could only be imposed where the land holdings formed part of a single estate, or where adjacent landowners agreed to co-operate, and where there was plenty of arable land. In well-wooded areas with small, scattered settlements, scope for the adoption of the open field system was limited.

The open field system reached its greatest extent in the twelfth and thirteenth centuries, but at all stages in its long history there were instances of piecemeal enclosure, as individual farmers withdrew their strips from the communal system. Existing field boundaries that have originated as a result of early enclosure of **medieval** open field strips are often characterised by a curving boundary like a reversed and elongated letter 'S'. This reflects the shape of the strip, where the curve at each end reduced the amount of space needed to turn when teams of oxen were used for ploughing. Since little new strip cultivation began after 1400, 'S'-shaped hedges or walls provide evidence that the boundary has been in existence since at least that time. At Laxton in Nottingham, the medieval strip field system is still in a limited form of communal use, and at Chelmorton in Derbyshire there are curved stone walls that enclose former strips. The regular corrugations of **ridges and furrows** visible in places in old pastureland are also evidence of former strip cultivation, the ridges having been formed as the result of a shift of soil towards the middle during each circuit of the strips by the plough. Occasionally, the larger earth banks or **baulks,** which were left unploughed between groups of strips or between fields, also remain.

Some enclosed fields dating from the medieval period originated as **assarts**, areas reclaimed from woodland under licence from the larger landowners. Assarts were therefore usually bounded by woodland edge, and the hedges probably constructed or thickened with saplings taken from the wood. Evidence that an existing hedge originated in this way can be provided by the presence of many plant species characteristic of local woodland.

The initial decline of the open field system coincided with the decline in population at the time of the **Black Death** in the fourteenth century, and was probably caused at least in part by a shortage of labour to cultivate the fields for crops. In time, whole estates were withdrawn from the system and the fields enclosed to accommodate sheep, which provided a more profitable form of land use. As demand for agricultural produce grew, and improved farming methods were introduced, there was increased pressure from the more progressive farmers and larger landowners to enclose land held in open fields. The first **Acts of Enclosure,** granting permission for the enclosure of open field parishes, were passed in the seventeenth century, but the most extensive phase of enclosure occurred between 1750 and 1850. It was usually a condition of enclosure that the external boundary of each new holding should be marked by a ditch and hedge, or a wall, within a year. Hedges and walls dating from this period tend to form a regular pattern of rectangular fields, and the hedges usually contain only one or two species, predominantly hawthorn. In the uplands of northern England and in Scotland, the main period of enclosure did not begin until the nineteenth century, so that the majority of Lakeland and Scottish walls date from this time. The fields established at that time were relatively large, and so have been less susceptible to change in recent years than those further south. The last major Enclosure Act was passed in 1903 for Skipwith in Yorkshire.

Since the Acts of Enclosure, economic forces have continued to influence field patterns. Cheap imports of food from overseas, towards the end of the nineteenth century, marked the beginning of a prolonged period of agricultural depression, only temporarily alleviated by the demands of the First World War. For some time, therefore, the established field pattern remained relatively unchanged. The food shortages of the Second World War stimulated a post-war drive towards more intensive farming systems, with increasing specialisation and widespread removal of boundary features as fields were enlarged, especially in the arable areas of the eastern side of the country.

In analysing field patterns as they are now, landscape historians distinguish between **planned countryside** and **ancient countryside**. The main area of planned countryside forms a diagonal wedge across England, centred on the Midlands and extending south to Wiltshire and north to Yorkshire (Figure 1.8). In general, boundaries in areas of planned countryside, such as the east Midlands, are most likely to date from the enclosure period of the eighteenth and nineteenth centuries, when the pattern of regular boundaries was superimposed on formerly open fields. The oldest field boundaries are most likely to be found in so-called ancient countryside, such as Devon, south-west Wiltshire, Herefordshire, the Welsh borders and lowlands, Essex and Kent, where a long history of small, dispersed and enclosed settlements meant that the landscape was little affected by the open field system and subsequent Acts of Enclosure.

Although the distinction between ancient and planned countryside can provide a useful general guide to the likely age of boundary features, it cannot do justice to the range of variation that exists. The rate at which new field systems have been adopted in a particular locality, or whether they have been adopted at all, has depended on a great many factors, such as when the area was originally settled, how easy the soil was to work, patterns of land ownership and inheritance, population density and the proportion of woodland, pasture and arable land. Even in counties such as Lincolnshire, Yorkshire, Norfolk and Northamptonshire, where eighteenth- and nineteenth-century enclosure was most widespread, less than 50% of the land area was affected by Acts of Enclosure. Very old boundary features are particularly likely to survive at the edge of holdings and along roadsides, especially where these coincide with parish boundaries.

While the existing pattern of field boundaries is very much a reflection of their past history, their structure is usually determined by their function, by

15

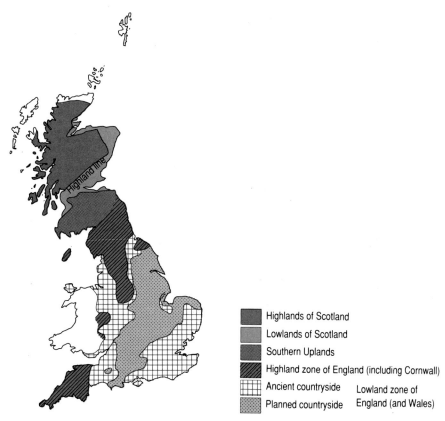

	Highlands of Scotland
	Lowlands of Scotland
	Southern Uplands
	Highland zone of England (including Cornwall)
	Ancient countryside Lowland zone of
	Planned countryside England (and Wales)

Figure 1.8 Map of the major regional landscape zones in Great Britain, showing the approximate extent of ancient and planned countryside in England and parts of Wales. (Source: Rackham, 1986)

the materials that are locally available and by the way in which they have been managed. Major regional differences in the distribution of different types of field boundary in England and Wales are shown in Figure 1.9.

Roads

Where prehistoric field systems still exist, as in many parts of Cornwall, it is likely that many of the roads between them are equally old. **Ridgeways** also are often of ancient origin, built along high ground to avoid the more densely wooded valleys, marshes and river crossings.

Many of the straight, long-distance routes that the Romans built for military purposes are still clearly evident and well known, but throughout Roman, **Saxon** and medieval times a network of minor roads and tracks developed to link fields and settlements. Most lanes and paths with a name are likely to date from at least Saxon times. Lanes that are sunk well below the level of the surrounding fields or that lie between steep-sided banks are also likely to be of considerable age. In some cases, large, double banks enclosing a ditch or lane were constructed to mark land ownership or administrative boundaries. In other places, especially on soft ground or where a lane runs down a steep hill, the lane has become hollowed out as a result of centuries of erosion.

Figure 1.9   Regional distribution of boundary features in England and Wales, 1985 (kilometres per square kilometre of non-urban land). (Source: Countryside Commission, 1990)

17

Drove roads originated as grassy tracks used mainly to drive stock between pastures, and in some cases to move them great distances from their home grazing areas to the markets in large centres of population. The network of drove roads was probably most extensively developed from Tudor times onward, but incorporated tracks that had been in use for centuries before this. They continued to be regularly used until well into the nineteenth century. Three pine trees planted alongside the road were used as a sign that overnight grazing was available, and in places these small groups of trees can still be seen.

Turnpike roads were introduced in the late seventeenth century, when turnpike trusts were set up to help maintain and improve major routes. The trusts could levy tolls and acquire land for new roads. Some old roads fell into disuse at this time and now exist only as footpaths or green lanes. Other roads were abandoned, and sometimes new ones were created, as a result of the reorganisation of fields during the time of the parliamentary enclosures.

Many roadside verges first became permanent features with the introduction of tarmacadam surfacing in the early nineteenth century. Surfacing meant that carriages, which had previously skirted mud and potholes, could be constrained to a narrower track, allowing vegetation to colonise the roadside. When surfacing was extended to minor roads in the 1930s, cost considerations meant that little-used roads were left to revert to green lanes. Since the 1950s, the trunk road network has been enlarged and the motorway system developed.

The inherited network of minor roads, green lanes, trackways and paths now adds interest to the landscape, and allows the public to explore the hidden recesses of the countryside on foot, bicycle, horseback or by car. Where the verges are relatively free from disturbance, they support a variety of wildlife, and in places where they have escaped fertiliser and pesticide use, they may support rich plant communities typical of old pastureland.

Canals

The majority of canals were constructed between 1758 and 1805. They were built to provide a cheap means of transport for raw materials and manufactured goods, and so are particularly common in the industrial Midlands. At its peak, the network of navigable rivers and canals extended to a length of 6800 kilometres. Although now little used for commercial purposes, canals add considerable visual interest to the landscape, especially where there are few natural expanses of water, and they frequently have great wildlife value. There is public access to the 2400 kilometres of towpaths that are under the management of British Waterways.

Railways

The canal transport system was quickly eclipsed by rail transport. Most railway lines were built between 1840 and 1870. As with canals, the aim was to keep gradients along the route to the minimum, and this involved massive earth-moving operations in the construction of cuttings, tunnels, embankments and viaducts. Some landowners refused to allow a railway line to cross their land, or insisted on conditions, such as the planting of screening avenues of trees, being fulfilled before granting permission.

Since the cutbacks of the 1960s, the length of operational railway track has been reduced by one-third. Disused lines have either been sold or simply been abandoned after removal of the rails. Both operational and disused railway lines have considerable wildlife value, especially if the vegetation is sympathetically managed.

1.4   The case study farm

The case study farm, which will be used throughout the book as an example of the management of boundary habitats, was chosen because it includes a wide variety of boundary features, and because the landowner has a particular interest in wildlife and so has experimented with several novel forms of field margins. He has had expert help with the identification and monitoring of the diversity of wildlife on the farm.

The Kemerton Estate covers 368 hectares and lies in south-east Worcestershire, close to the border with Gloucestershire. There are three different farming enterprises: arable on 262 hectares, a suckler herd of 45 red poll cows and a seasonal flock of 120 ewes on 61 hectares of permanent pasture. Woodland, a sand and gravel quarry, and buildings take up the remaining area.

Field sizes are relatively small compared with other farms in the area, with an average of 6 hectares for fields in the arable rotation and 2.5 hectares for the permanent pasture. This results in a correspondingly large number of field boundaries, with a total length of 21 500 metres.

The farm lies on the south-east flanks of Bredon Hill, an outlier of the Cotswolds, which rises from the Vale of Evesham. From the highest point on the farm, which rises to 155 metres above sea level, there are spectacular views over the Severn Vale into Gloucestershire. The highest ground lies on **oolitic limestone**, which has been used to build some of the field boundaries and many of the buildings. The lower slopes of the hill and the lower ground (down to 20 metres above sea level) lie over heavy **Lias** clays, which give way to **alluvial** gravels near the River Avon. Field drainage is important on the heavy clay soils, so many of the predominantly hawthorn hedges have ditches alongside. **Pollarded** willows, a characteristic feature of the area, occur in many of the hedges on the lower ground.

The farm lies on the edge of the Cotswolds **Area of Outstanding Natural Beauty** (AONB). The footpaths on the farm, particularly on the side of Bredon Hill, are popular with walkers from the local villages, and at weekends with walkers from further afield. There are four villages around the farm, and much of the farm situated on the flanks of the hill is visually prominent for people in these villages and on the roads between them.

The landowner has built up the farm from 142 hectares inherited from his father. Conservation has been an important feature of the farm's management for nearly ten years. The landowner has always had a special enthusiasm for bird conservation, but this has developed into a wider interest in natural history and, in particular, plants. The present farm manager came to Kemerton eight years ago. His main objective is to improve the economics of the farming enterprises, but he is also keen to manage the farm responsibly, with care for the natural infrastructure, including the hedges, ditches and woodland.

The farm therefore already has considerable landscape and wildlife value, but a conservation adviser from the county Farming and Wildlife Advisory Group has been called in to asess the scope for further improvement.

Chapter 2
LANDSCAPE ASSESSMENT

Boundaries such as hedges and stone walls feature strongly in many rural landscapes, and often play a significant part in defining regional character, the 'sense of place' which distinguishes one area of countryside from another. Even if they are not particularly distinctive, boundaries may be familiar and admired components of the local scene. They can also provide structure and interest in areas that are otherwise relatively flat and uninteresting, and soften areas dominated by unattractive buildings. When planning the management of boundaries, you should therefore give some thought to the way in which they relate to the landscape as a whole, as well as to their detailed appearance and the contribution that they make to the appearance of your land holding. Even if you are only planning the management of a single boundary feature, for example the renovation of a neglected stretch of hedge, it is worth first considering its landscape value and how this might change if you altered its management.

The main purpose of landscape assessment should be to identify those boundary features on a holding that have most landscape value and need to be maintained, those that show signs of deterioration and could be improved and areas where the creation of new boundary features might enhance the landscape. The assessment can be summarised as notes on a map, supplemented if possible by a report and photographs or sketches.

2.1   General landscape assessment

The character of a landscape is determined partly by its physical features – the landform, the vegetation and artificial structures – and partly by less tangible aspects, such as its cultural associations and the needs it fulfils. Landscape assessment therefore involves both *recording* what you see and *analysing* your feelings about it. Since landscape is a shared asset, you also need to give some thought to how other people might view the landscape that you manage.

Detailed guidance on the way to carry out a general landscape assessment is given in Chapter 2 of the foundation book. In summary, the main stages involved are as follows.

1   Select key vantage points on or overlooking the land holding concerned.

2   Record the components of the landscape from these viewpoints, in terms of the landform, main types of vegetation and structures.

3   Identify any clearly distinctive zones.

4   Identify the principal landscape features, i.e. those that are particularly notable or prominent.

5   Analyse how you feel about the landscape and the particular features in it, and if possible seek the views of others.

6   Add any available background information on previous management and points of historical or cultural interest.

Obvious vantage points to choose on farmland might include those on high ground on the holding, those giving a view up to or out of the farmhouse and those from which the farm is visible to others, such as public footpaths and roads. To assess the landscape value of transport routes, such as footpaths, bridleways, roads, railways and canals, journeys along the routes will be needed to establish the main landscape zones and features. As with farmland, views both from within and overlooking an area should be considered.

The amount of detail that you include in your assessment will depend on your purpose and the time that you have available. An example of a simple landscape assessment in the form of sketch maps is shown in Figure 2.1, in which the principal features are highlighted by descriptive notes. Different boundary types can be represented on a map using symbols (as in Figure 2.1), by letters (for example TH/T for tall hedge with trees) or by different coloured pens. For a full assessment, greater detail can be added in a written report, as in the case study example at the end of the chapter. Photographs and sketches can be valuable supplements to the assessment, and provide a useful basis for monitoring future landscape change.

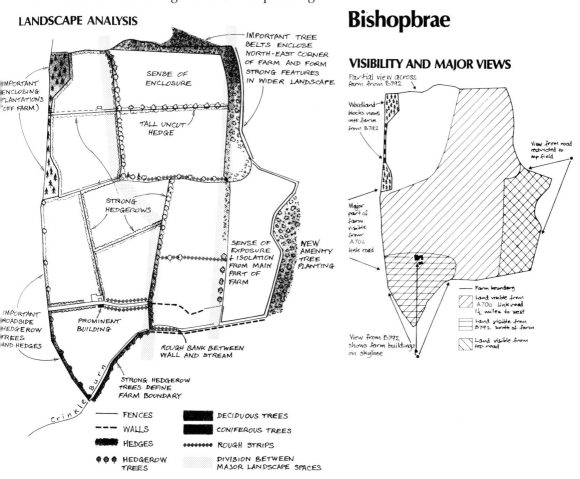

Figure 2.1   Example of landscape assessment in the form of annotated maps.
(Source: Cobham Resource Consultants, 1985)

Checklists, such as that shown in Table 2.1 of the foundation book, can help to ensure that no significant components of the landscape are overlooked. Your general assessment may reveal distinctive landscape zones, for example open landscape with stone walls on a hillside and smaller fields bounded by hedges and trees in the valley below. If so, you should make a note of this, as the approach to management of the different zones is likely to differ. Vantage points, distinctive zones and principal features should be marked on a map of the area; copies of an Ordnance Survey map on a scale of 1:10 000 or, for a small farm, a scale of 1:2500 are useful for this purpose.

Landscape assessment inevitably involves an element of personal judgement, so your records should include notes on how you feel about what you see. Do you consider it attractive or unattractive, and why? Table 2.2 in the foundation book provides some descriptive terms that may help you to put your feelings into words. It is important to consider, and if possible ask, how neighbours and passers-by view the landscape and feel about it, particularly where the land you manage is publicly visible. Some areas are considered to have such great scenic value that they have been officially designated as **National Parks**, Areas of Outstanding Natural Beauty, **National Scenic Areas** (NSAs) or **Environmentally Sensitive Areas** (ESAs); if any of your land falls into this category it should be recorded in your notes.

2.2   Assessing the contribution of boundary features

In a general landscape assessment, the main points to note about boundaries relate to their prominence in the overall scene (see the checklist in Table 2.1). For a more detailed description of the contribution that boundary features make to the landscape, the main points to consider are:

▶   the pattern that the features form;

▶   their structure;

▶   their historical and cultural significance;

▶   colours, smells, sounds and textures;

▶   your own feelings, and those of others;

▶   possible sites for new features.

Pattern

On farmland, notice the size and shape of the fields outlined by the boundaries, and how this pattern relates to the landform and to the field pattern in the area as a whole. Are the fields larger or smaller than those in the surrounding area? Are they rectangular or irregular in shape? Some idea of the great variation in field patterns that can occur has already been shown in Figure 1.7. Prominent boundary features, such as hedges and walls, will probably impose a clearly defined pattern, whereas fences, low banks and ditches on their own will have less visual impact, especially on flat land, unless they separate different crops or occupy a lot of ground. You should record any obvious variation in land use adjoining the boundaries, since this is likely to be an important part of the pattern, for example where a boundary separates moorland from improved pasture, pasture from arable land, cereals from oilseed rape.

Table 2.1 Checklist for landscape assessment of boundary features (coarse-grained features only; subjective assessment)

Hedgerows			
Average field size	More than 20 ha	10–20 ha	5–10 ha	Less than 5 ha
Continuity	No gaps	Few gaps	Several gaps	Many gaps
Height	More than 3 m	2–3 m	1–2 m	Less than 1 m
Standard trees	Many	Several	Few	None
Dry stone walls			
Average field size	More than 20 ha	10–20 ha	5–10 ha	Less than 5 ha
Continuity	No gaps	Few gaps	Several gaps	Many gaps
Height	More than 3 m	2–3 m	1–2 m	Less than 1 m
Transport verges			
Slope	Bank sloping away	Flat verge	Cutting rising up	
Width	More than 10 m	5–10 m	2–5 m	Less than 2m
Vegetation	Trees, shrubs, grassland	Shrubs and grassland	Grassland	

Notice also whether the boundaries are regularly or irregularly spaced, whether they follow a straight line or curve and whether they are continuous or gappy. Do they link up or embrace other significant landscape and wildlife features in the area, such as woods and ponds, or are these areas isolated from one another? Do any of the boundary features draw your eyes to a landmark, obscure a potentially attractive view or help hide an eyesore?

As a general rule, the pattern imposed by boundaries is likely to be more pleasing to the eye if it is in keeping with local character and with the scale and contours of the landform. A patchwork of small fields with high hedges and trees may enhance the intimate feeling of small settlements nestling in valley bottoms, but could look out of place where the beauty of an area lies in wide, uninterrupted views and skyscapes, as on the Salisbury Plain around Stonehenge. An angular pattern of square and rectangular fields may be attractive in a wide, straight river valley, but detract from the appearance of a gently curving landform of rolling hills and winding streams.

Gappy or discontinuous boundary features may be less pleasing than a continuous network, by giving an impression of being disjointed or incomplete. Tracks and roads may be visually intrusive unless they follow the lie of the land and other linear landscape features.

Structure

Once you have looked at the pattern of the boundary features as a whole, you can also look in more detail at their individual structures. For each boundary or boundary type, record the different features (for example bank, ditch and verge) that are associated with it. In the case of hedges, note the height, width, shape, density or gappiness, orientation and any evidence of distinctive management such as **laying** or **coppicing**. Previous management by laying is revealed by the presence of horizontal main stems, and coppicing by multiple stems arising just above ground level from a single

stool. If there are hedgerow trees, note their size (height and trunk diameter), frequency and any signs that they have previously been coppiced or pollarded. Examples of some of the variations to be found in hedgerow structure are shown in Figure 2.2. For banks, verges and cuttings, note the height, width, slope and aspect as well as the visual characteristics of any vegetation. If there is a ditch, note its width and depth, the presence or absence of water, whether the water is flowing or still, and any distinctive associated vegetation, such as tall reeds or bankside willows. Note the characteristics of any farm track, path or other area between any crop and the boundary. Along footpaths, tracks and lanes, record the presence of any unusual, old or locally characteristic features such as **squeezers**, kissing gates and milestones.

Tall hawthorn hedge

Laid hedge

Low hedge with mature trees

Gappy hedge

Figure 2.2   Variations in hedgerow structure

For a stone wall, note the height, width and length of the wall, the size range, arrangement and colour of the stones, and the colour of any obvious vegetation, such as **lichens**, growing on the wall. Note the state of repair of the wall, whether mortar or earth is involved in its construction and the presence of features such as sheep gaps (**cripple holes**), rabbit traps or drainage holes (**smoots**) and particularly large stones (**mearestones**, often used to mark the limits of a property or parish and sometimes inscribed with the landowner's name). Some of the regional variations in the structure of stone walls are shown in Box 2.1.

Box 2.1 Regional variations in the structure of stone walls

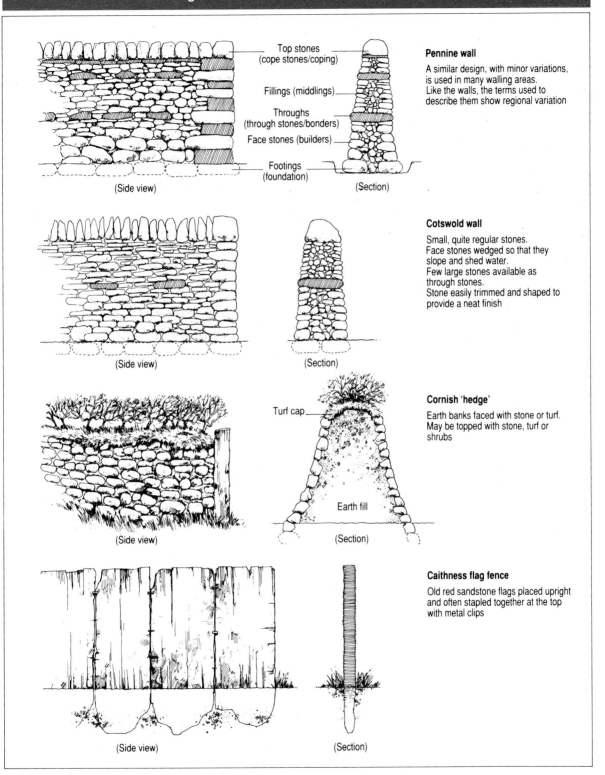

Pennine wall

A similar design, with minor variations, is used in many walling areas.
Like the walls, the terms used to describe them show regional variation

Top stones (cope stones/coping)

Fillings (middlings)

Throughs (through stones/bonders)

Face stones (builders)

Footings (foundation)

(Side view)　　(Section)

Cotswold wall

Small, quite regular stones.
Face stones wedged so that they slope and shed water.
Few large stones available as through stones.
Stone easily trimmed and shaped to provide a neat finish

(Side view)　　(Section)

Cornish 'hedge'

Earth banks faced with stone or turf.
May be topped with stone, turf or shrubs

Turf cap

Earth fill

(Side view)　　(Section)

Caithness flag fence

Old red sandstone flags placed upright and often stapled together at the top with metal clips

(Side view)　　(Section)

(continues overleaf)

Box 2.1 (continued)

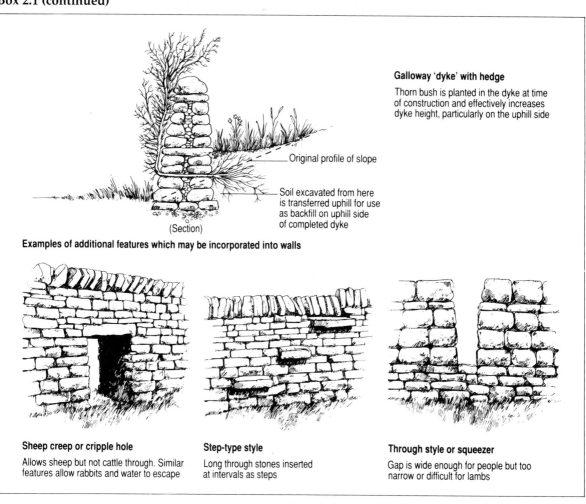

Galloway 'dyke' with hedge

Thorn bush is planted in the dyke at time of construction and effectively increases dyke height, particularly on the uphill side

Original profile of slope

Soil excavated from here is transferred uphill for use as backfill on uphill side of completed dyke

(Section)

Examples of additional features which may be incorporated into walls

Sheep creep or cripple hole

Allows sheep but not cattle through. Similar features allow rabbits and water to escape

Step-type style

Long through stones inserted at intervals as steps

Through style or squeezer

Gap is wide enough for people but too narrow or difficult for lambs

Historical and cultural significance

As mentioned in Chapter 1, many boundary features have a long and interesting history. It is worth trying to establish if this applies to any of your boundaries, since if so this will add to their landscape importance. Old photographs, maps and documents provide the most reliable sources of information. Old landscape paintings are also a useful, if more subjective, resource. Aerial photographs, dating mainly from the late 1940s onwards, provide a record of more recent changes; information about the aerial survey material that is available can be obtained from the Royal Commission on the Historical Monuments of England, the Royal Commission on the Ancient and Historical Monuments of Scotland and the Royal Commission on Ancient and Historical Monuments in Wales (see the *Helpful Organisations* supplementary booklet in the foundation module). Your county record office or planning department may be able to help you with documentation if you would like to follow up the history of the boundaries on your holding in greater detail.

The shape of field boundaries and the size of the fields that they enclose can provide additional clues to their origins, as mentioned earlier. Although less reliable, the characteristics of the vegetation can also provide a rough guide to the age and origin of a hedge. Some shrubs and trees, such as field maple, hazel, dogwood, service tree and small-leaved lime, and herbaceous plants, such as dog's mercury, bluebell and primrose, are characteristic of woodland, so the presence of these species in a hedge suggests that it may have been derived from woodland edge, possibly between the twelfth and fourteenth centuries. The method is not infallible, since under favourable conditions some of these species occur more widely, for example dog's mercury is common throughout Derbyshire.

The diversity of shrubs in a hedge can serve as another rough guide to its age. As a rule of thumb, if you count the number of different shrubby species in a 30-metre stretch and multiply that number by 100, you will have an approximate estimate of the age of the hedge (give or take 200 years!), at least where single-species hedges were planted originally. This method also has serious limitations, for example it is not always easy to distinguish between different hedge species, and the number of possible species varies from place to place according to soil conditions and climate. However, in very general terms, the more different shrub and tree species that a hedge contains, the older it is likely to be.

The size of any hedgerow trees can also provide clues to the age of a hedge. Most larger trees in an open situation grow in girth by approximately 2.5 centimetres per year, and smaller hedgerow trees (crab apple, pear, holly, hawthorn, rowan, whitebeam) grow by approximately 1.3 centimetres per year. **Diameter at breast height** (dbh) is an indication of the age of a tree and therefore of the minimum age of the hedge. The diameter can be measured either by using specially calibrated 'diameter tapes' or by using an ordinary tape measure to measure the circumference of the tree and dividing the figure obtained by 3.14.

In addition to their historical interest, some boundaries may have a particular significance for local people. For example, a secluded and little-used track may be known locally as lovers' lane or a roadside hedge as a good spot for blackberrying. It is useful to indicate in your notes the level of use of public footpaths and bridleways.

Colours, smells, sounds and textures

Do not forget the part that colours, smells, sounds and textures can play in your impression of the landscape, both in close and distant views.

Crops of different colour and texture emphasise field boundary patterns. Colourful flowers and berries in a hedge add to its visual interest. Try to visualise or recall the impact of seasonal changes – the red stems of dogwood in winter, the blossom of blackthorn and the young, green shoots of hawthorn early in spring, brown beech leaves in autumn – which all contribute to the visual interest of a hedge. The honey colour of Cotswold stone walls, the red sandstone walls found in Devon and eastern Scotland, the off-white walls of the Derbyshire Peak District and the dark slate used to bound fields in central and north Wales contrast attractively with the vegetation, and all make a contribution to colour in the landscape.

The sweet smell of honeysuckle in a hedge is likely to add to your appreciation of a landscape, and a ditch heavily polluted with effluent will probably detract from it. Bird song and the hum of insects foraging for nectar can add to the comforting and secluded appeal of a hedged track. Equally, the silence of stone-walled moorland, broken only by the call of a curlew, can contribute to an impression of remoteness and peace.

Your own feelings, and those of others

Remember to record how the landscape affects you. Do you find wide, open spaces exhilarating, threatening or boring? Do you find small fields with high hedges intimate and comforting or fussy and claustrophobic? Opinions about the same landscape can differ, and memories of familiar childhood landscapes can strongly influence our views. If you depend on the land for a living, your attitude may be dominated by commercial and managerial considerations. A boundary strip of bright red poppies alongside a crop of green-blue wheat may be greatly admired by passers-by, but may be seen as a threat to the crop by neighbouring farmers. Try to consider how other people might view your land, especially where it is overlooked, or visible from roads and paths.

It is worth 'sounding out' the opinions of neighbours, or at least keeping them informed, before making dramatic changes to boundary features. In areas designated as being of scenic interest, such as Environmentally Sensitive Areas, the contribution that boundary features make to the landscape is often acknowledged by grants for specified forms of maintenance, so any such designation should be recorded in your assessment. Occasionally hedgerow trees with particular significance to a locality are covered by a **Tree Preservation Order** (TPO: see the *Legislation and Regulations* supplementary booklet in the foundation module).

If you are responsible for the management of vegetation along transport routes, you need to consider the view of travellers as well as the view of the route from outside the boundary. Variation in the speed of travel along different types of route will affect how much travellers are likely to notice. For example, along footpaths and canals, fine details of the landscape will be visible, whereas along motorways and railway lines, only a general impression of the landscape and its most distinctive features will normally be gained. This can serve as a guide to the level of detail to include in your assessment, as well as to the scale likely to be needed for new features to be noticeable to travellers.

Possible sites for new features

If you are planning to introduce new boundary features, this should be kept in mind during your landscape assessment. From a landscape point of view, the most obvious places for siting new features, such as hedges and stone walls, are likely to be where there is a break in the boundary line, where investigation reveals that a boundary has been removed in the past or where new boundary features would link up isolated features or divide up fields that are uncharacteristically large for the locality or farm. Attractive new boundary features will have greatest landscape value if they are sited where they are publicly visible, for example alongside public footpaths and tracks. On arable land, new sites that run north–south are likely to be more acceptable for hedge planting than those that run east–west, because hedges aligned north–south cast less shade on the crop. As in the case of existing boundary features, new boundaries are more likely to have wide appeal if they are constructed of local materials, conform to a traditional local pattern and structure, and are in sympathy with the contours of the land.

2.3　Carrying out your own landscape assessment

The previous sections have discussed the main points to consider in carrying out a landscape assessment, but the best way to gain confidence in the subject is by putting the ideas into practice. If you do not own or manage any land, you could try out the techniques on a publicly accessible area such as a roadside verge or country park. At first you may find it difficult to describe a landscape that you have previously taken for granted; if this is the case, try to imagine that you are describing the scene to a blind person.

You will need:

▶ Ordnance Survey maps (or sketch maps based on these) showing your area and some of the surrounding countryside;
▶ pens and paper for making notes and sketches;
▶ checklists such as those shown in Table 2.2, and in Tables 2.1 and 2.2 in the foundation book;
▶ a camera and binoculars if you have them.

For a *general* landscape assessment, follow the stages summarised in Section 2.1. Mark and number your selected vantage points on a map. Record the components of the landscape that are visible from these points, as well as any clearly distinctive landscape zones and notable features on the holding. On transport routes, travel along the route to identify the main characteristics. If possible, take photographs or make sketches of the views from the vantage points and of any particularly interesting features. Make brief notes on your reasons for selecting the viewpoints and on your general impressions of the landscape. Add to these notes any background information that you are able to obtain about the area and its history. Where applicable, include information on public access and any statutory designations.

For an assessment of *boundary features*:

▶ mark their position on a map, using symbols, codes or coloured pens to distinguish between different types;
▶ add notes to the map to highlight any stretches that are particularly prominent or important features in the landscape;
▶ make additional notes to record more detail on the visual characteristics of each boundary type, your personal perceptions and those of others with an interest in the area, and any background information on historical or cultural associations.

2.4　Landscape assessment of boundaries on the case study farm

The following landscape assessment of the case study farm, carried out by a conservation adviser on behalf of the landowner, can be used as a guide to carrying out your own assessment. Remember, though, that any assessment is a personal view. The purpose and scope of your assessment, the distinctive landscape features and your own perceptions may well differ from this

example. If so, do not be afraid to modify the approach to suit your needs, provided that all the main points are covered.

Kemerton Estate, Worcestershire

Background information

The geology and the resulting topography of the land has created an attractive and in places striking landscape. Lying on the flanks of Bredon Hill, the farm is divided into two roughly equal halves, each with its own landscape character. The northern area lies on the slopes of the hill on limestone and middle series Lias clay, while the southern area lies in the flatter vale over lower Lias clay and alluvial gravels.

Field sizes over the whole farm are relatively small, giving rise to the type of gentle 'patchwork quilt' views often considered typical of rural England. Most of the fields around the village of Kemerton and on the steeper slopes of the hill are permanent pasture, the rest being arable. Until the arrival of **Dutch elm disease** in the early 1970s, hedgerow elms were a dominant force in the landscape. The landscape assessment map (Figure 2.3) shows the distribution of hedges with elm **suckers** currently growing in them, all of which formerly supported elm trees. However, despite Dutch elm disease, the farm still has an intimate, wooded appearance. This is largely due to the ash trees that have been allowed to grow up through many of the hedges, the pollarded willows growing along the ditches, the small blocks of woodland and the small orchards around the villages (plum orchards were a mainstay of the local economy until 30 years ago). Tall hedges and young plantations have been used to screen a sand and gravel quarry on the estate. Old maps of the farm dating from the **Enclosure Awards** of 1813–38 show that there has been remarkably little change since then in the pattern of field boundaries.

There are about 8 kilometres of footpaths on the farm, with 2.5 kilometres running along farm tracks, 2 kilometres through grass fields and 3.5 kilometres in arable fields, most running around the outside edges. Several of the footpaths are used regularly by local people, and two on the higher ground on Bredon Hill are popular with walkers from further afield. The roads running between the villages of Bredon, Kemerton, Kinsham and Westmancote are all well used and give good views across the farm, particularly the higher parts of the farm on the slopes of Bredon Hill.

Overall impression

The overall impression is of a typical English rural landscape with small fields and mixed land use. The network of hedges and small blocks of woodland are the dominant features, and pollarded willows are noticeable in places on the farm. Many of the hedges are trimmed low, except around permanent pasture on the slopes of Bredon Hill, but hedgerow trees and the small blocks of woodland give a warm, intimate feel to the landscape.

Landscape zones

In terms of landscape, the farm can be divided into two main zones: the sloping ground on the flanks of Bredon Hill to the north and the flat land of the vale to the south. The part of the farm on Bredon Hill is highly visible from the villages of Bredon, Kemerton, Kinsham and Westmancote – indeed, the slopes of the hill can be said to dominate the villages. The part of the farm on the flatter land in the vale has fields that are similar in size and land use to those on the hillside. However, from within this zone, views are restricted by the network of hedges and hedgerow trees, as well as by the blocks of woodland and the small orchards in the villages, creating a more intimate and enclosed landscape.

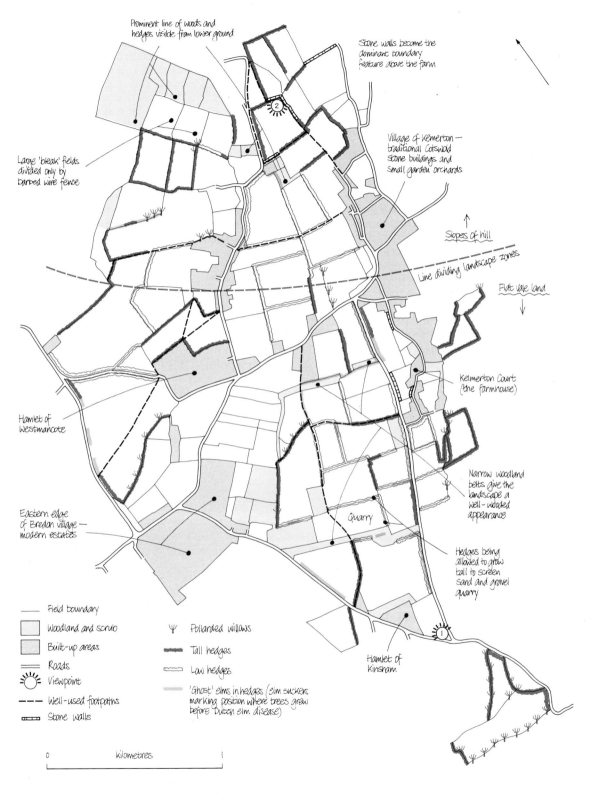

Prominent line of woods and hedges visible from lower ground

Stone walls become the dominant boundary feature above the farm

Village of Kemerton — traditional Cotswold stone buildings and small garden orchards

Large 'break' fields divided only by barbed wire fence

Slopes of hill

Line dividing landscape zones

Flat vale land

Kemerton Court (the farmhouse)

Hamlet of Westmancote

Narrow woodland belts give the landscape a well-wooded appearance

Eastern edge of Bredon village — modern estates

Quarry

Hedges being allowed to grow tall to screen sand and gravel quarry

Field boundary

Woodland and scrub

Built-up areas

Roads

Viewpoint

Well-used footpaths

Stone walls

Pollarded willows

Tall hedges

Low hedges

'Ghost' elms in hedges (elm suckers marking position where trees grew before Dutch elm disease)

Hamlet of Kinsham

0   kilometres   1

Figure 2.3   Landscape assessment map for Kemerton Estate

31

Two viewpoints were chosen to encompass the main features of each zone: one looking up at the slopes of Bredon Hill and one from the highest point on the farm looking down over the vale. Their positions are marked on the map shown in Figure 2.3.

Viewpoint 1

Viewpoint 1, on the road from Kinsham to Kemerton, is typical of many in the vale. A photograph and an annotated sketch of the view from this point are shown in Figure 2.4. Little can be seen of the fields in the vale, since the hedgerow trees shut out views across the fields. However, looking up towards Bredon Hill, most of the fields and field boundaries on the northern half of the farm are visible. Undulations in the hill slope give the mainly rectangular fields more shape. The strong network of hedges and blocks of woodland create an attractive patchwork pattern, emphasised by the changing colours of the various crops as they develop and flower, and the different textures of soil through the year. Two large, grass fields stand out as rather bleaker than their surroundings. The north–south hedges are less obvious in the landscape than those running from east to west. One line of strong hedges and woodland blocks running across the slope is particularly prominent.

Viewpoint 2

Viewpoint 2, from the highest point on the farm, gives striking views over the Vale of Evesham. Apart from a field in the foreground, most of the land towards the village of Kemerton is hidden by a small spinney and tall hedges. Beyond this, several areas with a more wooded appearance lie amongst the small fields. As from viewpoint 1, it is the hedges and fields running from east to west that are dominant. With the countryside stretching out over mixed farmland towards the Cotswold escarpment, the view from this point on the farm can be truly breathtaking on a clear day.

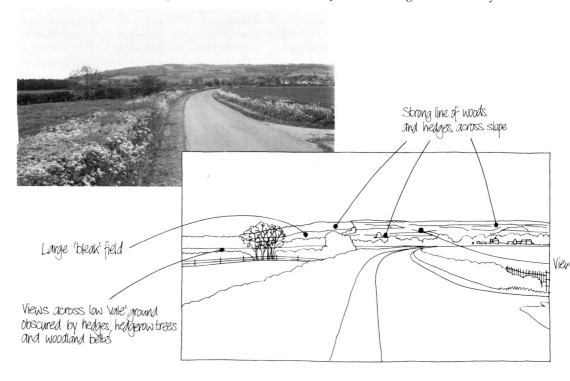

Strong line of woods and hedges across slope

Large 'break' field

Views across low 'vale' ground obscured by hedges, hedgerow trees and woodland belts

View

Chapter 3
HABITAT ASSESSMENT

The wildlife interest of boundary features depends on the interaction of human activity with environmental factors (such as soil and climate) and ecological processes (such as dispersal, colonisation, competition and **succession**). Although the relative contribution of these factors varies from site to site, most boundary features exist only because of human efforts and needs, and survive only because of regular, careful maintenance. Conversely, the wildlife value of boundary features is susceptible to damage by neglect, and by disturbance from intensive use of adjoining land.

An appreciation of the way in which natural factors and management practices influence the distribution of wildlife will enable you to recognise the boundaries likely to be of most conservation value. It will also serve as a useful guide to the selection of appropriate management options for the future.

This chapter provides examples of the way in which natural factors affect the wildlife value of boundary habitats. It discusses the use of four, simple, habitat assessment criteria, and illustrates with reference to the case study farm how to carry out a boundary habitat assessment.

3.1 Species distribution and diversity

Environmental factors

Environmental factors that influence the distribution of wildlife, particularly plants, include soil type, climate and structure. These natural factors are frequently modified by human activity in the creation and management of boundary features.

Soil type

Most hedgerow shrubs tolerate a wide range of soil conditions, but a few are particularly sensitive to the presence or absence of calcium. Dogwood, the wayfaring tree and purging buckthorn, for example, are all **calcicoles** or lime-loving plants, found mainly on **alkaline** soils, such as soils on chalk or limestone. In general, boundary features on **neutral** or alkaline soils are likely to develop a richer flora than those on **acid** soils.

The same principle also applies to stone walls; limestone walls, such as those made of Kentish ragstone or Cotswold stone, support more species than walls composed of hard, acid rocks, such as gritstone. Where lime mortar has been used in the construction of a wall, it can allow the growth of lime-loving types of lichens, mosses and ferns, even if the stone in the wall is non-**calcareous**.

In a few cases, the soil type has a direct influence on the distribution of animal species too: snails, for example, need lime for their shells and so are more common on lime-rich soils. More usually though, the influence of soil type is indirect, through its effects on food supply (since the larvae of glow-worms feed on snails, they too occur mainly on alkaline soils).

Plant diversity is affected by soil fertility, less fertile soils generally being more species rich. Fertiliser drift and run-off encourage the development of aggressive, fast-growing species (such as coarse grasses, nettle and bramble), which then dominate and crowd out less competitive species (such as fine-leaved grasses and many herbaceous wild flowers). Most field margins and roadside verges have vegetation characteristic of high soil fertility. Sites that have never received fertiliser and that have developed a rich and stable plant community are usually highly valued in conservation terms.

Fertiliser run-off can harm the wildlife interest of water-filled ditches, the rich nutrient supply encouraging the growth of **algal blooms** and blanket weed, which smother and shade out other aquatic vegetation and deplete the water of oxygen as they decay.

Climate

Climate can influence wildlife distribution by both its macro and its micro effects. On a large scale, for example, fewer hedgerow and tree species, and breeding birds, are found in the generally harsher climatic conditions in northern England and Scotland than further south. On a small scale, small, localised variations in moisture and light, which determine the micro-climate, can have a marked influence on the distribution of some plants and animals. South-facing stone walls, fully exposed to the sun, may be too dry for any plants except the most drought-tolerant lichens. On the other hand, north-facing walls, especially in the wetter west of the country, may remain relatively humid and damp and so support the growth of mosses and ferns. Cracks in a wall, ledges and upright, unmortared **copings** can trap sufficient moisture to allow the growth of **xerophytes** (drought-tolerant plants, such as the succulent-leaved stonecrops) where conditions are otherwise too dry. Small gradations in moisture content may also affect the distribution of plants (and some animals) on banks and verges; the upper zone of a bank is likely to be drier than the lower zone and, since roads need to be well drained, the outer, roadside zone of a verge is likely to be better drained than the inner zone. In addition, the roadside zone may be slightly warmer because of heat reflected from the road.

Hedges that cast a very deep shade, such as beech, prevent the establishment of other plants at their bases. However, moderate shade allows the survival in hedges of plants more characteristic of woodland habitats, since they will be free from the competition of faster-growing, sun-loving plants. In general, sites with a sunny, southerly aspect are likely to have a greater diversity of wildlife than those facing north, except where there are other limiting factors, such as drought stress, as in the case of stone walls. Warm, sunny, dry walls and banks are especially good habitats for cold-blooded terrestial animals, such as insects, spiders, lizards and snakes. The ballast on railway lines can similarly act as a suntrap for reptiles. In ditches and canals, underwater light levels will affect the distribution of aquatic plant and animal life. Water courses that are heavily shaded, or turbid as a result of pollution, are likely to support less aquatic wildlife.

Structure

A boundary feature will be particularly valuable as a habitat for animals if its structure is varied enough to meet a number of needs. Examples of the way in which various aspects of the structure of a hedge meet the nesting and feeding requirements of different birds are shown in Table 3.1. Tall trees in hedges provide songposts for birds such as the yellowhammer, thrushes and blackbird, and nest sites for crows, magpie and woodpeckers. The presence of trees in a hedge benefits most hedgerow birds, despite the fact that trees also provide lookout posts for nest predators. Hawthorn hedges

Table 3.1 The hedge as a habitat for birds*

Upper branches of hedgerow trees

Nest only	Nest and feed	Feed only
Carrion crow	Wood pigeon	Blue tit
Rook	Greenfinch	Chaffinch and other tree and shrub species
(Buzzard)	(Treecreeper)	
(Kestrel)		
(Mistle thrush)	(Lesser spotted woodpecker)	

Trunk and holes

Nest only	Nest and feed	Feed only
Barn owl	Wren (in ivy)	Treecreeper
Little owl	(Green woodpecker)	
Stock dove	(Greater spotted woodpecker)	
Jackdaw	(Nuthatch)	
Great tit		
Blue tit		
Starling		
Tree sparrow		
(House sparrow)		
(Kestrel)		
(Tawny owl)		

Shrubs

Nest only	Nest and feed	Feed only
Turtle dove	Wood pigeon	Fieldfare
Magpie	Cuckoo (lays in hedge sparrows' nests, etc.)	Redwing
(Collared dove)	Long-tailed tit	
	Song thrush	Mistle thrush
	Blackbird	Robin
	Lesser whitethroat	Great tit
	Hedge sparrow	Marsh tit
	Wren	Blue tit
	Goldfinch	Whitethroat
		(Willow tit)

Shrubs continued

Nest only	Nest and feed	Feed only
	Linnet	(Coal tit)
	Lesser redpoll	(Willow warbler)
	Bullfinch	(Chiffchaff)
	Chaffinch	plus
	Greenfinch	(Pied flycatcher, redstart and other species on migration)
	(Moorhen (especially over ditches))	
	(Blackcap)	
	(Garden warbler)	
	(Red-backed shrike)	
	(Cirl bunting)	

Herbs, low brambles

Nest only	Nest and feed	Feed only
	Whitethroat	Goldfinch
	Yellowhammer	Greenfinch and other shrub and ground species
	Reed bunting	
	(Nightingale)	
	(Chiffchaff)	
	(Grasshopper warbler)	
	(Sedge warbler)	
	(Stonechat)	

Ground

Nest only	Nest and feed	Feed only
Skylark	Robin	Hedge sparrow
	Corn bunting	Blackbird
	Pheasant	Song thrush
	Partridge	Wren and many other shrub and herb species
	Red-legged partridge	
	(Willow warbler)	

(Source: Slightly modified from Pollard et al., 1974)

*Species listed in brackets are less commonly found, either because they are rare or because the hedge is a less favoured place for them.

with straggly outgrowths often provide a profusion of readily accessible berries, which are particularly attractive to blackbird, redwing and fieldfare in winter. Dense hedges, and those with plenty of dry grass at the base, provide nesting cover for birds such as the whitethroat, goldfinch and dunnock. No single hedge type is likely to meet all these needs, so diversity in the management of hedges on a holding will add to their habitat value.

In the case of ditches, variations in the water depth, rate of water flow and substrate all contribute to habitat diversity. Greatest diversity is likely in deep, water-filled ditches with gently sloping banks. In such situations, the gradation in water depth can allow a large variety of plant types to thrive. In many ditches, structural and therefore habitat diversity is limited, but diversity can be increased by small variations in the structure of different stretches.

Many animals that spend much of their time on or near water are encouraged by the presence of other boundary features in addition to water. Dragonflies ideally need a broad band of swamp vegetation, in an area exposed to the sun but backed by sheltering woodland; amphibians, such as the frog and toad, need cover near water for hibernation sites and foraging grounds. Similarly, few terrestrial animals can meet all their requirements for food, breeding sites and shelter within a single feature, such as a hedge, and most require access to water. The close proximity of a number of boundary features, such as a ditch, hedge, bank and track, will therefore increase the habitat value of a boundary.

Ecological processes

Plants

Ecological processes, such as dispersal, colonisation, competition and succession, play a significant part in determining the distribution and diversity of the plant life associated with boundary features. The effects of these processes can be illustrated by examining the origins of species-rich hedges.

Most hedges established since the time of the Acts of Enclosure, at least from Victorian times onwards, were probably planted with a single species, usually hawthorn since this rapidly produces a dense, thorny, stockproof barrier. Occasionally, other species have been used, for example beech on the edge of Exmoor, *Fuchsia* in north-east Scotland and the Channel Islands, holly in parts of Staffordshire, Scots pine in Breckland, and *Pittosporum* or *Escallonia* around the daffodil fields of the Isles of Scilly. Georgian enclosure hedges often included a second species, usually ash, elm or hazel. These were planted to provide a dual-purpose hedge, serving both as a barrier and as a source of timber when coppiced.

Species-rich hedges, which contain more than one or two woody species, may have arisen in a number of ways. They may be old assart boundaries, derived from a mixture of woody plants at the woodland edge; they may occur where hedges that were originally planted with only one or two species have slowly become colonised by other shrubby species; they may have developed naturally along undisturbed boundaries, from wind- or bird-dispersed seeds; or they may have been planted deliberately as mixed hedges.

Once a hedge is established, the dense growth makes natural colonisation by other species a slow and uncertain process. The number and type of shrubby species in a hedge can therefore provide an indication of its age, as mentioned in Chapter 2. Successful, long-term colonisation depends on a nearby source of seed, the right conditions for germination and growth, resilience to repeated trimming, and longevity. Elder, for example, is often found as a colonist in hedgerows, because it is readily dispersed by birds, establishes in existing hedges where the ground has been disturbed (for example by badger setts) and flourishes with hard cutting back. However, it is not a very long-lived species and is often removed during management before hedges are laid, so it is not a useful indicator of an old hedge. Rose is moderately successful as a hedgerow colonist. Like elder, it is spread by birds and is unharmed by regular trimming. It is often left when hedges are laid, and occurs in both relatively new and old hedges. Hazel, field maple and dog-wood, on the other hand, do not readily colonise an existing hedge, and their presence suggests that a hedge is at least 400 years old and was established at a time when woodland cover and nearby sources of seed or saplings were plentiful. Spindle is characteristic of hedges that are at least 600 years old. However, these species also occasionally occur in more recent hedges where the necessary conditions for their colonisation and survival still exist, or where they have been planted deliberately. For example, field maple and dogwood are common in hedges in parts of Wiltshire.

In some cases, the number of shrubby species in a hedge can decline with age, as a result of uneven competition. For example, species that sucker readily, such as elm, can take over in a mixed-species hedge. This process can be accelerated by a change in management that encourages one species more than another. For instance, regular coppicing may formerly have kept elm in mixed hedges in check, whereas nowadays management by routine mechanical trimming, which treats elm in the same way as the rest of the hedge, allows it to sucker and spread freely.

In the same way, the processes of dispersal, colonisation and competition affect the type of herbaceous plant community that becomes established in hedge bottoms, banks, field margins and verges. Bare patches of soil created by cultivation, herbicide sprays, fire, heavy grazing or traffic damage are rapidly colonised by fast-growing **annuals** (often considered to be perni-cious weeds) from the soil seedbank or other nearby seed sources. Slower-growing **perennials** take longer to establish and spread, but gradually form a dense cover. This makes it more difficult for annuals, which only survive the winter as seed, to re-establish, unless bare soil is once again exposed.

Along roadside verges, saline conditions caused by the run-off from salting roads to prevent ice forming in winter can kill salt-sensitive plants, so creating bare patches. More salt-tolerant species are then able to colonise this ground. Accelerated dispersal by traffic has allowed some maritime plants (such as reflexed salt-marsh-grass, sea plantain and sea aster) to spread considerable distances inland to exploit this new roadside habitat.

The way in which natural and human influences, combined with ecological processes, affect the vegetation along railway lines is illustrated in Box 3.1.

Well-maintained hedges, verges and ditches gradually acquire a rich and stable **flora** and **fauna** characteristic of their type. Without regular manage-ment, succession to a new habitat type would occur. Ditches would silt up and aquatic species become replaced by terrestial ones, the grasses and other herbaceous plants of banks and verges would gradually be replaced by scrub, and hedges would develop into a gappy line of trees, which would eventually become over-mature and die out.

Railways are unique habitats, which result from a combination of environmental and human factors, including drought stress, annual herbicide spraying, cutting and burning regimes, and train-assisted and human-mediated seed dispersal. Examples of the way in which these factors favour the growth of particular species are given below.

Drought stress

The exposed nature of a railway track, with its free-draining foundation of ballast, favours plants that are adapted to prolonged periods of drought. These include annual species, such as whitlow grass, small toadflax, ratstail fescue and rue-leaved saxifrage, which can lie dormant as seeds until conditions are right. Other well-adapted species, such as field horsetail, adders tongue fern, meadow saxifrage and purple toadflax, have deep-rooted stolons (fleshy bases to the stem), characteristics that also allow these plants to survive burning.

Annual herbicide spraying

The track ballast is sprayed with persistent herbicide every summer, with the aim of keeping the track free of vegetation. However, some annuals seem resistant to low concentrations of herbicide and are able to grow later in the season as the effects of the herbicide wear off; examples include whitlow grass, Canadian fleabane and American willowherb. The perennial meadow saxifrage flowers and dies back before spraying, surviving underground in the form of bulbs.

Bank mowing

Most railway banks used to be cut on an annual basis and the hay collected for horse fodder. Coarser material was raked into rows after drying and burnt, which kept nutrient levels low, encouraging fine fescue grasses, the ubiquitous false oat-grass and a herb-rich sward. Herbs still common on chalk embankments in southern England include field scabious, small scabious, agrimony and greater knapweed, the latter often with its associated parasite, knapweed broomrape.

Bank burning

Before the advent of diesel and electrically powered trains, railway bank vegetation was regularly burnt, either as part of a regime of cutting and controlled burning, or in accidental fires caused by the discharge of hot coals and sparks from steam locomotives. Plants that flourished in burnt areas included rose-bay willowherb (also known as fireman's lupin), the onion-rooted variety of false oat-grass, field horsetail and adders tongue fern.

Train-assisted seed dispersal

Many seeds, particularly the parachute seeds of the composite family, are swept along the rail corridor by the vortex of the train. A well-documented example is Oxford ragwort. Other species introduced in this way, which have spread rapidly, are Canadian fleabane and American willowherb, already mentioned as problem weeds for British Rail because of their resistance to the herbicides normally used on the track. Occasionally, rare species, which have been dispersed by falling onto the lineside from goods wagons, are found, having been brought into the country as contaminants of hay or wool. One plant showing this form of dispersal is a Balkan trefoil, *Trifolium pannonicum*, which has only been discovered at two locations on the Marylebone line in Buckinghamshire.

Human-mediated seed dispersal

Railwaymen have traditionally grown garden plants around their stations, signal boxes and huts. Some of the plants still exist on the banks close to where they were originally planted, long after the plots have been abandoned. These include *Iris* and rose species, lupin and privet. Others that were well adapted to the railway environment, such as everlasting pea with its deep-rooted stolons, the butterfly bush, Japanese knotweed and biting stonecrop, have not only survived but spread.

Animals

The value of boundary habitats for animal wildlife will be dependent on their richness as a source of food as well as on the degree to which they can provide breeding sites and shelter from weather, predators and disturbance. Since all animals depend either directly or indirectly on plants for food, boundary features that have an extensive plant community are likely to attract a varied and numerous animal population. The area occupied by the boundary, and its form of management, will also influence its suitability as a source of food supply to animals.

The mammals that are commonly found in hedgerows are shown in Table 3.2. These mammals and their food sources are only part of a series of interlinked **food chains** that exist among the animals associated with hedgerows. For example, woodmice feed on seeds in the hedgerow and in turn form part of the diet of owls:

Seeds ➔ Woodmice ➔ Owls

Each part of this food chain has other links. Seeds are eaten by many other creatures, such as shrews and finches; woodmice are preyed on by hawks, foxes, stoats and weasels as well as owls; owls take prey other than woodmice. In reality, therefore, the food chain is part of a complex network of food relationships or **food web**.

At each successive stage in a food chain, part of the energy originally built up in the plant tissue at the base of the chain is lost. Predators at the top of a food chain are therefore dependent on plentiful sources of their prey in order to meet their energy needs, and their presence indicates an area rich in wildlife. Those with a restricted diet may require large stretches of suitable habitat in order to survive. For example, barn owls require at least 2 kilometres of grassy margins alongside hedges and ditches as a flight path in order to seek out their prey.

Table 3.2 Common hedgerow mammals and their food sources

Species	Habitat	Food sources
Red fox	Burrows (earths)	Rabbits, rodents, birds, large insects (e.g. beetles), earthworms, carrion, berries
Badger	Burrows (setts)	Rabbits, rodents, earthworms, much vegetable material (e.g. plant roots, acorns)
Rabbit	Burrows (warrens)	Any green vegetation, particularly young shoots
Stoat	Vegetation	Rodents, young rabbits, sitting birds
Weasel	Vegetation	Rodents, young rabbits, sitting birds; can climb trees
Hedgehog	Vegetation	Insects, earthworms, slugs
Mole	Burrows	Insect larvae and other arthropods, earthworms
Common rat	Burrows	Earthworms, insects, molluscs, vegetable material (e.g. grain)
Field vole	Vegetation and tunnels	Grasses and sedges
Common vole	Vegetation and tunnels	Grasses and sedges
Bank vole	Vegetation and tunnels	Variety of soft vegetation
Woodmouse	Vegetation	Seeds of all kinds
Common shrew	Vegetation	Insects, spiders, woodlice, snails, some plant material (e.g. seeds)
Pygmy shrew	Vegetation	Insects, spiders, woodlice, snails, some plant material (e.g. seeds)

(Source: Adapted from Dowdeswell, 1987)

As mentioned in the foundation book (Chapter 3 and Figure 3.3), plant species that have been present in the country for a long time and are widely distributed often provide food for a much greater variety of insect life than non-native species; as a result they also provide indirect support for more insectivorous (insect-eating) species. Of the trees associated with boundary features, willow and oak are particularly valuable, each capable of supporting well over 400 different insect and mite species. The common hedgerow shrubs are also rich in insects and mites, more than 200 species being associated with hawthorn and more than 150 with blackthorn. Of the herbaceous species commonly found in boundaries, stinging nettle has 27 insects closely associated with it and a further 17 insects that use it in their diet; these include the caterpillars of the small tortoiseshell, red admiral, peacock, comma and painted lady butterflies. Grasses too support a large insect population, including several butterfly species.

Other plants, while supporting fewer species, may nevertheless be valuable as the main or only food source for a particular species. For example, the distribution of the brimstone butterfly depends primarily on the occurrence of purging buckthorn and alder buckthorn. Information on woodland and hedgerow butterflies and their main food plants is given in Table 5.4 of *Woodlands,* another of the books in the *Practical Conservation* series.

The availability of breeding sites and shelter in boundary areas will further affect their habitat value for animals. Requirements for breeding success are often very specific. To take the example of the grey partridge, a ground-nesting species, favoured nesting sites are gappy hedge bottoms where there is plenty of dead grass cover to prevent detection by predators, and hedges growing on banks are preferred. If such ground-nesting birds are disturbed, for example by inquisitive dogs, their breeding success can be reduced. Grey partridge chicks forage away from the hedge, at first on insects and then on weed seeds. Unsprayed crop margins or **conservation headlands** adjoining a boundary therefore provide them with the food and cover they need to survive (see Chapter 6).

In general, the greater the structural diversity of boundary features and the larger the area that they cover, the greater the range and number of potential breeding sites that they will offer.

3.2 Criteria for assessing the wildlife value of boundary habitats

Once you have acquired an understanding of the main factors that influence the distribution and diversity of wildlife, as described in Section 3.1, you will have a good basis for assessing the value of a boundary as a habitat. To some extent this will be a matter of personal judgement, since there is no truly objective or exact measure. There will always be situations in which even experts will disagree, or cases where personal preferences have an important influence, but the following criteria are a useful guide to the wildlife value of a boundary:

▶ its naturalness or origin;
▶ the diversity of species that it contains;
▶ the rarity of the species;
▶ its area or size.

Naturalness or origin

Whilst few boundary features can be described as truly natural, the longer that they have been in existence, the more likely they are to contain a rich variety of plants, which in turn provide food and shelter for a large number of animal species. Naturalness in the case of boundary features therefore relates mainly to their origin or age. Even though the nature of a boundary may have changed over the years, for example where a hedge has been planted or developed naturally on what was once just a bank, the vegetation is less likely to have been disturbed than if it had been under cultivation, and a more varied flora will probably have survived. An example of the way in which the criterion of origin can be applied to hedges is given in Table 3.3. A similar example for water-filled ditches is given in *Water and Wetlands*.

Naturalness also involves some consideration of the extent to which a feature has been protected from damage due to intensive use and management of the adjacent land. Herbicides, pesticides, fertilisers, excessive grazing, careless burning, compaction and disturbance by traffic, and run-off from the salting of roads can all reduce the wildlife value of boundary habitats that otherwise have high potential value because of their age. If applicable, such information should be recorded in your notes.

Diversity of species

In assessing the wildlife value of boundary habitats, you need to consider both the abundance and diversity of wildlife species present and the overall structural diversity (since this will affect the variety of micro-habitats available for colonisation).

Species diversity is most easily assessed by recording the number of different plants (especially woody plants in the case of a hedge) associated with the feature, since all animals are dependent on plants for their survival, either directly or indirectly through the food chain. If you have time, you can also record the more obvious animal species, such as birds and butterflies. For these, bear in mind seasonal differences and the fact that there is likely to be some natural variation from year to year. For example, insects, small mammals and birds are particularly vulnerable to harsh winters and shortages of food.

Remember that soil and climate can affect species diversity, so that comparisons between sites are only valid where these influences are similar. In order to monitor changes in species diversity over a period of time, you need to use a standard procedure. For example, for hedges record the number of different shrub and tree species, or birds or butterflies seen, in a representative 30-metre stretch. You will need to make several visits if you want a complete record: plants are most easily identified when they are flowering or seeding; the presence of birds is most noticeable when they are nesting; butterflies are most active on warm, sunny days. It is not essential to be able to identify everything by name, provided that you have some way of distinguishing between different species. However, you will find it more interesting and informative if you can recognise the wildlife that you are recording. Some of the many excellent illustrated guides that are available to help with identification are listed in Appendix I.

To record structural diversity, you need to note such things as the presence of trees in a hedge, whether a hedge is associated with a bank and ditch, and variations in depth in ditches. Table 3.3 indicates how the plant species and structural diversity of hedgerows can be assessed.

41

Table 3.3 Habitat assessment criteria for hedges

Criterion	Rating
1 Origin	
Relict ancient woodland (assart hedge)	****
Ancient boundary hedge (pre 1700)	***
Enclosure hedgerow (post 1700)	**
Recently planted (post 1900)	*
2 Vegetation diversity	
2.1 Woody plant species diversity	
More than 10 native tree and shrub species	****
Between 6 and 10 native tree and shrub species	***
Between 2 and 5 native tree and shrub species	**
Single native species	*
2.2 Structural diversity	
Hedge with standard trees, shrubs and wide hedge bottom (more than 1 m either side)	****
Hedge with no standard trees but wide hedge bottom (more than 1 m either side) *or* hedge with standard trees, shrubs but little hedge bottom (less than 1 m either side)	***
Hedge with no standard trees but a thick base (laid in the last 20 years)	**
Hedge with no standard trees and a thin base	*
3 Plant rarity	
Containing one or more native plant species that is nationally or regionally rare	****
Containing one or more native plant species that is locally rare or unusual (e.g. found nowhere else on the farm)	***
Containing mainly common native plant species	**
Containing only non-native species	*
4 Size	
4.1 Volume	
Hedge more than 3 m high and more than 2 m wide (= 6 m³)	****
Hedge 2–3 m high and 1.5–2 m wide at base (= 3–6 m³)	***
Hedge 1–2 m high and 1–1.5 m wide at base (= 1–3 m³)	**
Hedge less than 1 m high and less than 1 m wide at base (= 0–1 m³)	*
4.2 Connectivity and continuity	
Interconnected network with no gaps	****
Interconnected network with few gaps (less than 10% of length)	***
Network with many gaps (10–20% of length)	**
Network with substantial gaps (more than 20% of length)	*

Rarity of species

Rarity may apply to either plant or animal species but, as in the case of diversity, plants are easier to record. A species may be rare in the country as a whole, or in a particular area. It may be restricted to a few individuals wherever it occurs, or be relatively abundant but only in certain areas.

Rare species may depend for their survival on both a rare habitat type and a particular form of management. For example, the tiny rush *Juncus mutabilis* is a rarity that grows on cart tracks on moorland on **loess** soils and depends for its survival on very occasional use of the track. The way in which a hedgerow can be assessed for plant rarity is shown in Table 3.3.

Size

The greater the area covered by a boundary feature, the more wildlife it will be able to support. Wide boundary features, for example wide grass verges and railway banks or green lanes with broad hedges on either side, offer wildlife greater protection from the effects of adjoining land use. The greater the length of the feature, the more scope there is for structural diversity, for example a ditch may be narrower or deeper in some parts than others.

Height or depth will be important in the case of three-dimensional features, such as hedges, banks, walls and ditches, allowing species with slightly differing habitat needs to co-exist in the same feature at different levels without serious competition. For songbirds, the volume of a hedge affects its value as a habitat: a thick hedge of moderate height may be as valuable a habitat as a tall, thin one, possibly partly because of the greater protection that a thick hedge offers from predators.

In assessing boundaries on the basis of their size, consideration should also be given to their continuity, and the extent to which they form an interconnected network and link with other areas of semi-natural vegetation, such as woods. For some species, boundary features that form corridors between larger blocks of habitat are likely to be more valuable than those that are isolated. Table 3.3 indicates how a hedge can be assessed on the criterion of size.

Using the criteria

The simplest way to use these criteria is to assess each feature by comparison with others of the same type on your land or nearby. A star rating system can then be applied to record your assessment, as shown in Table 3.4 for some of the case study farm boundaries. If used singly, the criteria may not be a reliable guide to conservation value. For example, a single-species hedge is unlikely to be of exceptional conservation value, however great its size. Conversely, there may be considerable overlap between some of the criteria. For example, a hedge that scores highly in terms of its naturalness or origin will probably also have a high score for species diversity.

Habitats that score highly on more than one of the four criteria are likely to be of high conservation value. Your assessment should allow you to establish which boundaries on your holding have a high conservation value and should be maintained, which have a moderate conservation value and could be improved, and whether there are areas with little existing wildlife interest where the introduction of new features might be considered. The ratings for individual boundaries should provide you with ideas for how to improve them. For example, a low, gappy hedge could be improved by infilling the gaps or planting trees in them, and by encouraging the hedge to thicken at the base and grow taller.

3.3 Carrying out your own boundary habitat assessment

As with landscape assessment, assessing the wildlife value of boundaries is best done by marking important areas on a map, taking photographs or making sketches from recorded positions, and making detailed notes describing representative as well as significant stretches.

While carrying out your assessment, keep in mind the dynamic nature of the factors responsible for the existing wildlife value of the habitat, and the type of management that will be needed in future to maintain or improve it.

In a few cases, the wildlife of a boundary may already have been professionally surveyed, for example if it falls within a designated area such as a **Site of Special Scientific Interest** (SSSI), or if it is an old roadside hedge along a parish boundary. If so, the county council or local wildlife trust should have a record of the survey information and you may be able to use this to supplement your own assessment (see the *Helpful Organisations* supplementary booklet in the foundation module).

As for the landscape assessment, you will need maps, checklists, blank paper and pens, plus a camera and binoculars if you have them.

▶ Mark all the boundary features on a map, as for the landscape assessment.

▶ Select representative stretches of each similarly managed boundary type for detailed assessment. Complete a checklist for each boundary feature, as shown for hedges on the case study farm in Table 3.4, if necessary adding any notes on the reasons for your ratings. Summarise the main information with notes on the map, as shown for the case study farm in Figure 3.1.

▶ If you have time and an interest in wildlife identification, include in your assessment a brief report on the wildlife associated with each habitat type, as shown for the case study farm in Section 3.4. Photographs and sketches of significant features are helpful if your assessment will be used by other people, and will be useful as a basis for monitoring future progress; record on the map the vantage points from which they were taken.

3.4 Habitat assessment of boundaries on the case study farm

A variety of field boundary habitats are present, including hedgerows, field margins and trackways, ditches and streams, stone walls and wire fences. Those of particular wildlife value are noted on the habitat assessment map shown in Figure 3.1.

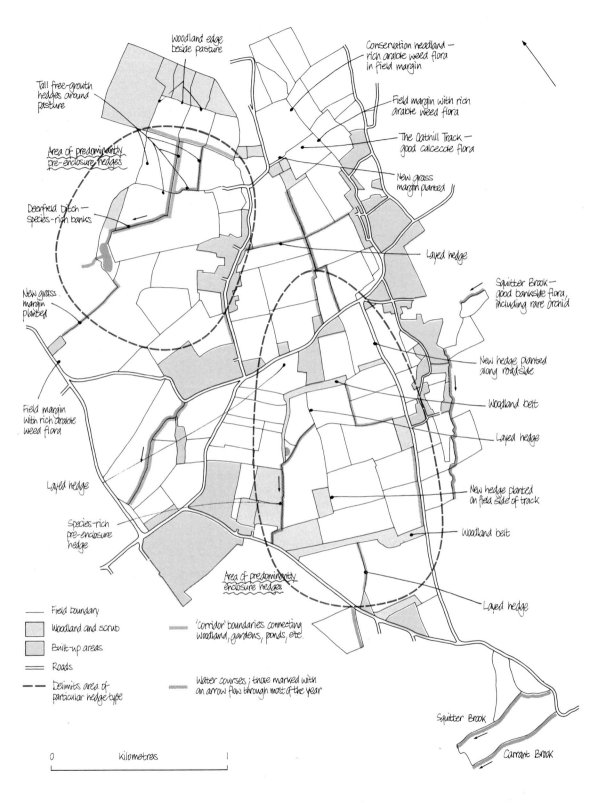

Figure 3.1   Habitat assessment map for Kemerton Estate

In the hedgerows, hawthorn, elm suckers, blackthorn, wild plum, field maple and dogwood are the dominant shrub species, while dog rose, bramble, black bryony, white bryony, hedge bindweed and old man's beard are the most common rambling species. Old man's beard dominates many hedges on the higher ground. Ash is the most common hedgerow tree, although oak, pollarded willow, poplar, horse chestnut, sycamore and wild plum are common along certain hedges.

Two distinct areas of hedge can be distinguished on the farm (see Figure 3.1). Those in the north-western part of the farm are dominated by blackthorn and all contain dogwood, while those in the southern part of the farm are dominated by hawthorn and include wild plum, but no dogwood, with the exception of one very species-rich hedge. Although soil type could be partly responsible for defining these two areas, the history of their owner-ship is also different, the two areas only having been amalgamated under the present owner.

It is likely that the blackthorn/hawthorn/dogwood hedges are pre-enclosure field boundaries, which ran beside drainage ditches in the medi-eval open field system. The hawthorn/wild plum hedges are probably more recent, originating during the eighteenth-century enclosures. This theory is reinforced by old maps of the area. The wild plum in the enclosure hedges has naturalised from the orchards that used to abound in the area.

The exception to this pattern is the species-rich hedge already mentioned. Although this hedge now lies within the enclosed area, it probably once formed the boundary of a large, medieval open field containing a commu-nally used free well. (One of the adjacent fields is called 'Frankwells', mean-ing 'free wells', and has a spring in it; old maps show several paths leading to one point in the field.) The hedge contains hawthorn, blackthorn, ash, dogwood, elder, field maple, hazel, spindle, sallow, willow, wych elm, English elm and honeysuckle. A habitat assessment checklist for the three principal hedgerow types (the two predominant types and the exception) is shown in Table 3.4.

Table 3.4 Habitat assessment checklist for hedges on the case study farm

		Enclosure hedges	Pre-enclosure hedges	Old medieval field boundary
1	Origin	✳ ✳	✳ ✳ ✳ assumed from map evidence and species compostion	✳ ✳ ✳ (possibly ✳ ✳ ✳ ✳)
2	Diversity			
	2.1 Woody plants	✳ ✳	✳ ✳	✳ ✳ ✳ ✳
	2.2 Structure	✳ to ✳ ✳	✳ ✳	✳ ✳ ✳
3	Plant rarity	✳ ✳ but contains 'wild plum' of agricultural origin from orchards	✳ ✳	✳ ✳ ✳
4	Size			
	4.1 Volume	✳ ✳	✳ ✳ ✳	✳ ✳ ✳ ✳ (abuts a wood)
	4.2 Continuity	✳ ✳ ✳	✳ ✳ ✳ ✳	✳ ✳ ✳ ✳

The bird life of the hedgerows is typical of that found in lowland England, as has been confirmed by a survey of breeding songbirds carried out on the farm in 1987 by the Royal Society for the Protection of Birds. Breeding wren, blackbird, song thrush, long-tailed tit and yellowhammer are all reasonably common in the hedges, as are the redwing and fieldfare in winter. Less common breeding birds found on the farm in thick hedges and scrubby field corners are the grasshopper warbler, whitethroat, lesser whitethroat and occasionally the nightingale.

The farm contains a number of small, isolated blocks of woodland, and it adjoins many gardens and small orchards in the surrounding villages. The habitat value of these wooded areas is considerably enhanced by the network of interlinking hedges. Certain boundaries are of key importance as 'corridors' for birds, insects and mammals travelling between patches of woodland, and these are marked on the habitat assessment map (Figure 3.1).

The field margins and trackways

The flora of the field margins (or hedge bottoms) varies according to the soil type, adjoining land use and past management practices. Where the soil overlies limestone and is light, the hedge bottoms contain the finer grasses, such as meadow-grasses and fescues, with a variety of wild flowers, such as greater knapweed, field scabious, agrimony and common St John's-wort. On the heavier clays, the coarser grasses, cocksfoot, tall fescue and false oat-grass, are dominant; there are generally fewer wild flowers, but hogweed, red campion, herb Robert, hedge mustard and black horehound occur.

In many field boundaries, previous disturbance, either physical from machinery, or chemical from herbicide and fertiliser drift, has favoured the more aggressive weed species over the finer grasses and wild flowers. Some years ago, this trend was exacerbated in some boundaries by the deliberate spraying out of weeds. Along several boundaries, mainly in arable fields, agricultural weeds such as barren brome, cleavers, couch grass and field bindweed have become dominant.

In general, the flora of the trackways shows the same variation as that of the hedge bottoms. However, because the width of a track gives protection against disturbance from the field, some tracks on limestone have become notably rich in species. The Oathill Track is one example, with 32 species of wild flowers recorded. During 1989, two new grass trackways were established beside arable fields. In time, these should develop a varied flora of perennials and become good corridors for wildlife, as well as giving access for crop inspection.

During the last four years, some field margins have been extended to create conservation headlands, with the occasional **fallowed headland** (an area cultivated but not cropped), to encourage certain types of wildlife on the farm. These headlands have been surveyed, and several species of annual plant that are now nationally rare have been recorded. These include prickly poppy, night-flowering catchfly, corn buttercup and cornflower, as well as locally rare annuals, such as Venus'-looking-glass, round-leaved fluellen and sharp-leaved fluellen. Field margins on the soils on the light limestone brash on Bredon Hill and the alluvial gravels in the vale support the most diverse annual floras, with up to 13 species per square metre being found.

Ditches and streams

There are over 3 kilometres of field ditches on the farm. However, most of these are dry during the summer and do not support a truly aquatic flora. Those that do, such as Deerfield Ditch and Carrant Brook, contain species such as reed sweet-grass, common reed, fool's watercress, hairy willowherb, water figwort and brooklime, and support aquatic insects, such as dragonflies like the banded agrion and the brown hawker. Fish, such as roach and eel, are also found in Carrant Brook.

One stream, Squitter Brook, which has had little engineering improvement and regularly floods onto the surrounding field edges, has an especially rich flora, including the rare green helleborine. This appeared on spoil when the stream was cleaned out a few years ago and was recorded nowhere else in the county at the time.

Stone walls

Although many of the buildings are made from locally quarried limestone, few of the field boundaries on the farm are stone walls. Conversely, off the farm, on the top of Bredon Hill, where sheep grazing was once common, and where limestone is directly at hand, stone walls are the usual form of field boundary. The walls are built in the Cotswolds style (see Box 2.1). Some of the older walls on the farm support lichens, although air pollution has reduced the variety of species that occur. Mosses, biting stonecrop and white stonecrop also grow on undisturbed walls.

Summary of the wildlife interest of the farm boundaries

With respect to wildlife, there are several field boundaries on the farm that are particularly diverse and contain a range of uncommon species. These are marked on the habitat assessment map shown in Figure 3.1. The ancient pre-enclosure hedges in the north-western part of the farm and the old, open field boundary hedge are of special value, as are most of the tall and wide hedges. Field edges and trackways still retaining an undisturbed flora, especially on limestone, such as the Oathill Track, should be conserved, while the problem, weed-dominated field edges found elsewhere are candidates for improvement.

Chapter 4
COMMERCIAL AND INTEGRATED ASSESSMENT

Assessment of the commercial value of a site will play an important part in formulating a management plan that will integrate conservation and commercial interests. Commercial considerations, such as the availability of labour, machinery and money, will usually need to be taken into account, even if conservation has a high priority.

On farmland, the scope for the conservation of boundary features will be affected by the nature and range of commercial enterprises, the type of stock kept and crops grown, the intensity of production, the number of people employed and the financial viability of the farm. Where boundary features border transport routes such as roads and railways, the safety of users will be a priority, and low-cost maintenance is also likely to be an important consideration. Legal obligations relating to the land may be involved in the assessment. For example, farm tenants may be required by their tenancy agreements to keep hedges well trimmed, public rights-of-way have to be maintained, the boundaries of motorways and railway lines must be kept stockproof. In some cases, especially where recreational land uses are involved, boundaries may contribute directly to the commercial value of a site.

Detailed advice on assessing the commercial value and potential of an area of land is beyond the scope of this book. For the purposes of developing an integrated management plan for boundary features, it is sufficient to concentrate on recording a number of key points in the commercial assessment:

▶ the main commercial enterprises and their extent;

▶ any subsidiary land uses or sources of income, for example shooting or fishing;

▶ the overall profitability of the land, the relative profitability of different enterprises and, equally important, the landowner's attitude to the existing levels of profitability;

▶ the resources of labour and machinery likely to be available for conservation work;

▶ external obligations or pressures concerning the land or enterprises, such as tenancy conditions, production quotas, markets, safety concerns.

The commercial assessment should be combined with landscape and habitat assessments to obtain an integrated assessment of the value of boundary features on the land. This may expose instances where the management of boundaries for commercial purposes and management to enhance their conservation value conflict to some degree. One of the main aims of an integrated assessment should be to identify ways in which any such conflicts might be minimised, and any potential beneficial interactions exploited to the full.

Attitudes and interests can play an important part in determining where the balance lies between what are seen as management problems and management opportunities. Landowners or farm managers with an interest in field sports or bird-watching, for example, may be more prepared to accept weeds in the field margin and crop edge than those with a purely commercial interest in the land. If you are a landowner, you will probably unconsciously incorporate your attitudes into your management plan. However, it may be worth trying to make your views explicit, in order to ensure that they have not blinded you to particular values and options, and to make your plan clearer to others who may need to consult it or carry it out. One way of doing this might be to seek out the opinions of others and see how they compare with your own. Consulting others, or at least keeping them informed of your plans, is also a good idea in terms of public relations, particularly if you are planning any changes to the management of boundary features that will affect the owners of adjoining houses or land. If you are an adviser, determining the attitudes and interests of the landowner and, where relevant, the manager of the land will be vital to the successful implementation of any management plan.

When making an integrated assessment, you need to give particular attention to features with a high conservation value that are adjacent to intensively managed land, and note any practices that may be affecting the wildlife interest of these features. Identify any areas that are not essential to the main land use, which might be managed in order to develop their conservation interest. Note any aspects of land use that appear susceptible to change. For example, on a farm, an enterprise that is losing money is likely to be discontinued or its management intensified or altered, which could have implications for the boundary features involved. Features that show signs of losing their landscape or wildlife interest through neglect or mismanagement also deserve special attention.

This chapter highlights the main interactions between the management of boundary features for commercial purposes and their management for conservation, and includes suggestions for minimising conflicts where these occur. With the help of the case study example, it shows how a commercial assessment can be combined with landscape and habitat assessments as the basis for an integrated management plan. Greater detail on the management options for conservation that are mentioned is provided in Chapters 5 and 6.

4.1 Interactions between the commercial and the conservation aspects of land use

Hedges and walls

Boundary features such as hedges and walls are most likely to have agricultural value if they form stockproof barriers on predominantly livestock farms, or if they provide shelter from the wind for high-value crops, such as early flowers and soft fruit. They may also be valued as farm boundaries, especially along roads, where they discourage trespass, and as shelter for gamebirds, where there is an interest in shooting. Other less easily quantified, but nevertheless important, agricultural benefits may result from reduced soil erosion on light land, shade and shelter for stock in extreme weather conditions, and improved pollination and pest control by insects that shelter and overwinter in hedges. Uncropped field margins allow easy access for farm vehicles and facilitate crop inspection.

Conflict between commercial and conservation objectives in the management of boundary features is most likely on intensively managed farms, particularly those that are all arable and where the landowner or farm manager has little interest in wildlife or game. In these circumstances, hedges may be perceived more as an agricultural liability than an asset, occupying potentially productive land, hindering the movement of large machines, harbouring weeds, pests and diseases, and costing money to maintain. However, most arable farmers take a pride in the appearance of their farms, and there are conservation measures that can be taken for which the savings outweigh any costs.

Stockproof barriers

On farms with livestock, where hedges or walls are valued as stockproof barriers, agricultural and conservation interests will probably coincide. Hedges can be allowed to grow up and need only be trimmed every two or three years, or occasionally rejuvenated by laying to maintain a gap-free structure. Except when fields are over-grazed, so that stock begin to damage the bark of bushes and erode the soil at their bases, damage to hedges from the management of adjacent land is far less likely than on arable land.

Shelter

In the case of certain high-value crops, such as nursery stock, soft fruit and early spring flowers, hedges or shelter-belts may be planted specifically for wind protection, especially in exposed situations. A hedge is very effective at reducing wind speed, because it forms a semi-permeable barrier. Whereas a solid barrier, such as a wall, blocks the path of the wind completely, forcing it upwards and creating turbulence, particularly on the leeward side, a hedge cushions the force of the wind, significantly reducing its speed downwind for a distance of up to 20 times the height of the hedge (Box 4.1). For complete wind protection, fields must be small (about one-quarter of a hectare), and bounded by tall hedges on all sides, as in the case of the daffodil fields on the Isles of Scilly. Where frosts are a problem, care is needed to site hedges so that they do not trap cold air. Less common hedging species, such as alder, poplar or birch, may be preferred for shelter, depending on the situation and the crop. They may have relatively little wildlife interest, since both they and the adjacent land are likely to be intensively managed, but may well be highly valued in landscape terms as a characteristic regional feature. Their habitat value can be improved by planting them in staggered rows and allowing a coarse grass strip to develop in the space between.

The sheltering effect of hedges is optimal when they are 40–50% solid, that is when people or stock moving on one side, but not standing still, can be seen indistinctly through the structure by someone on the other side. Tightly clipped hedges may be more dense than this in summer, and the resulting wind eddies can contribute to **lodging** (flattening of the crop) in cereals close to the leeward side of a hedge. Gaps in a hedge, or bare stems at the base, can increase rather than reduce wind damage, by funnelling the wind through at these points.

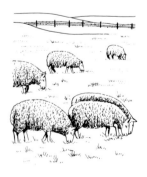

The extent to which hedges and walls have an economic value as shelter for stock is not clear cut. Although cattle and sheep unquestionably seek out and benefit from the comfort of a hedge in extreme weather, studies have found no difference in weight gain between sheltered and exposed stock.

Box 4.1 Hedges as shelter

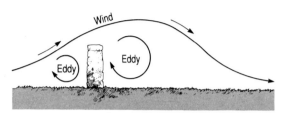

Impermeable barrier

A solid barrier, such as a wall, forces the wind upwards, creating eddies in front of and behind the barrier

Permeable barrier

A semi-permeable barrier, such as a hedge, allows the wind through, reducing its speed

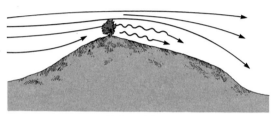

Hedge on hill crest

A hedge sited along the top of a ridge or on a hillside has its height effect enhanced by the slope of the land behind it, making it a more effective shelter when the wind is in the right quarter

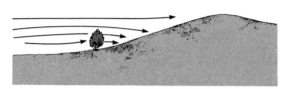

Hedge on slope facing the wind

The most difficult fields to protect are those that slope into the prevailing wind, which in most cases blows from the south-west. Here the effective height of the hedge is greatly reduced and on southerly slopes shading is detrimental to crops

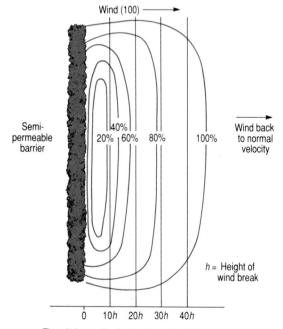

The wind speed is significantly reduced for up to 20 times the height of the hedge

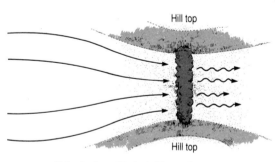

Valley hedge to block wind funnel (Top view)

Winds tend to be deflected or funnelled into valleys. Therefore hedges most effectively block these winds when located at right angles to the valley

Sheep may be protected from wind, rain and snow behind a tall hedge, but conversely may well be buried if snow drifts through it. However, in the uplands and especially in exposed areas, appropriately sited and well-managed hedges and shelter-belts can significantly improve conditions for livestock, reducing the occurrence of climate-related diseases such as pneumonia and twin-lamb disease. The slightly raised temperatures in well-sheltered fields can encourage early grass growth, allowing stock to be moved onto them sooner than more exposed sites.

Hedges and shelter-belts can be used to reduce wind-induced soil erosion on light soils. This can occur in dry years with crops, such as sugar beet, for which a fine seedbed is necessary. Shelter-belts of pine and willow have been planted in parts of the Fens to prevent the windblow of valuable peat soil and seed. The shelter-belts now form a characteristic and attractive regional feature, as well as habitat for wildlife. The willow can be cropped for firewood, pulp and craft purposes. *Soil erosion*

Walls, hedges and hedgerow trees can shade adjoining land to a distance of between one and two times their height. Where hedges are allowed to grow tall for conservation purposes, or contain many large trees, this can create problems of uneven ripening of crops and reduced yields at the field edge on arable farms. Hedgerow trees also shade hedges themselves, which can eventually lead to the development of gaps either side of trees and reduce the value of hedges as stockproof barriers. There are several ways in which this potential conflict between management for conservation and management for commercial purposes can be minimised. Taller hedges can be restricted to areas where there is a wide track or boundary strip on the shaded side, or the opportunity can be taken to create a new track in the shade. Hedges aligned north–south can be allowed to grow taller than those aligned east–west, since a north–south alignment allows sunlight to reach all the crop for at least part of the day. Taller stretches of hedge, or new hedgerow trees, can be restricted to the corners, or junctions, of hedges, which are particularly valuable areas for hedge-nesting birds. Small, deciduous trees, such as hazel, field maple, bird cherry, blackthorn and whitebeam, which will only cast light shade, can be selected as hedgerow trees. Trees that are late coming into leaf, such as ash, can be selected, although ash is shallow rooting, so that its roots may be damaged by ploughing if cultivation is carried up to the field edge. *Shade*

The labour and time needed for hedge maintenance can cause problems, particularly on arable farms devoted to autumn-sown crops, where access may be restricted once the crop is established. In this case, a less intensive trimming regime, in which hedges are cut every two or three years rather than annually, and only a proportion of the hedges tackled in any one year, or only one side trimmed each year, can be beneficial both for agriculture and for wildlife. If a wide boundary strip can be left between a hedge and the crop, this allows access for trimming after crops are sown and helps spread the workload, as well as allowing easy crop inspection at all times of year. Such a margin provides additional wildlife habitat, especially if sown with grasses and wild flowers, and protects the hedge from potentially damaging operations in the field. *Maintenance*

An alternative low-cost form of hedge management might be to coppice just above ground level once every ten years or so. Provided that only a proportion of the total length of hedge is coppiced in any one year (for example 10% each year for a ten-year coppice rotation), wildlife should be able to

53

recover from the disturbance that this management involves. Coppicing may have a dramatic, if temporary, effect on the landscape. However, where a hedge consists of a double or staggered row, this impact can be reduced by cutting only one side of the hedge at a time, if landscape is a particularly important consideration (for example close to a farmhouse).

Grants are available to help with the costs of improving existing hedges (for example by laying or coppicing), but not for annual maintenance (see Section 7.4 on grants in Chapter 7).

Stone walls, provided that they are well constructed and regularly checked for loose or missing **top stones**, may remain in good condition with minimal maintenance for well over 50 years. Although extensive repairs, if necessary, may be costly and require skilled labour, grant aid is often available (see Chapter 7). In the long run, it may be easier and cheaper to restore a damaged or derelict wall than to dismantle it and replace it with a fence.

Movement of machinery   Hedges and walls that enclose small fields can impede the movement of machinery, particularly if the boundaries are irregular in outline. Small and awkwardly shaped fields increase the time needed for manoeuvring machinery on the headlands, around obstacles and through gates. Optimum field size from an economic point of view depends on the size of the machine being used. Whatever the size of machine, the increase in work rate achieved by hedge removal to enlarge fields tails off with increasing field size (Figure 4.1).

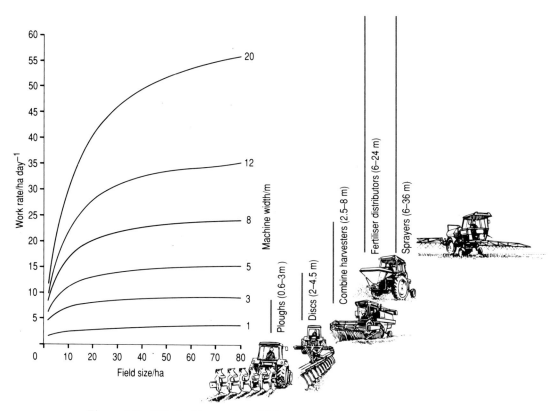

Figure 4.1   *Effect of increasing field size on work rate for machinery of different widths. (Source: Adapted from Sturrock and Cathie, 1980)*

54

Even if there is little interest in conservation or field sports on the part of the landowner, there may still be advantages to modest-sized fields. A large, uninterrupted expanse of open ground can be boring for the machinery operator. Existing boundaries may delineate fields of differing soil type requiring different cultivations and cropping patterns, best retained as separate units. The retention of existing boundary features and field sizes on arable farms allows scope for future diversification, and enhances the opportunities for recreational land use.

Awkward field corners might be developed as conservation areas by leaving them fallow or planting them with trees to create a more easily worked field shape, rather than by removing a hedge to improve access.

Large fields can be divided to improve their wildlife interest, with minimal obstruction to the movement of machinery, by the creation of low, grassy banks.

Weeds and the arable flora

Hedges and field margins are often blamed by farmers for harbouring the more troublesome agricultural weeds. In practice, although many agricultural weeds occur in field boundaries, research has shown that of these only barren brome and cleavers present any significant threat to the crop. Both can be kept in check in the boundary by cutting or mowing at the time of flowering to prevent seeds being formed and shed. Other grass weeds, such as rough meadow-grass, although common in field boundaries, rarely spread into the crop.

Accidental or deliberate spraying of hedge bottoms, and fertiliser drift, create conditions in which annual weeds can flourish. Where the field margin is undisturbed so that there is a dense, perennial grass sward at the base of the hedge, most problem, annual, arable weed species are crowded out, to the advantage of both agriculture and conservation. With a few notable exceptions, such as couch grass and common horsetail, most perennial weeds are unable to survive in annually cultivated fields.

The widespread use of herbicides and fertilisers, the change from spring to autumn drilling and improved methods of seed cleaning mean that some formerly serious arable weeds, such as corn crowfoot, corn cockle and shepherd's-needle, are now threatened with extinction. Novel forms of extended field margins are being promoted as a way of encouraging the survival of some of the more showy or unusual arable weeds.

Hedges and field margins may be viewed as a source of crop pests. Although some pests of non-cereal crops do overwinter in field margins, they are usually fairly easy to control and often specific to a narrow range of host plants. For example, the larvae of the *Brassica* pod midge feed on the developing pods of oilseed rape, but damage is often confined to the crop edge, and spray treatment, if justified, can be restricted to this area. The larvae of the pea and bean weevil feed on peas and spring-sown field beans, especially on the roots and root nodules. The adults overwinter in hedges and ditches, so damage is often most noticeable at the crop edge, but a range of insecticides is available for the control of outbreaks in the crop. Black bean aphid, which affects field beans, sugar beet and a range of horticultural crops, overwinters on spindle trees, but a forecasting system exists to help in its control, and sugar beet crops are normally protected by the sprays necessary to control peach potato aphid. Stem **nematode**, which attacks a wide range of species, can be spread from weeds common at the field edge, such as chickweed, cleavers and wild oat, but does not usually cause serious

Pests and beneficial insects

damage. Other pests associated with hedges include the damson hop aphid (a pest of hops that overwinters on blackthorn and plums), the lettuce root aphid (occurring on Lombardy poplar and lettuce) and the apple twig cutter, a weevil that overwinters in hedge bottoms.

Of greater significance may be the fact that some wild grasses, including the perennial rough meadow-grass, may harbour barley yellow dwarf virus (BYDV), which can be transmitted by aphids to cereal crops. Early-drilled winter cereals are at a susceptible stage when autumn aphid migrations occur. Autumn mowing of grassy field margins, and spray treatment of grassy crop stubbles in the field before ploughing, should reduce the presence of BYDV.

Many of the insects and spiders that shelter in field margins are beneficial from the agricultural point of view, attacking pest species or serving as pollinators for crops such as beans, oilseed rape and fruit trees (see Figure 4.2). Predators depend on their pest prey for survival, so that in some cases the presence of a small reservoir of pest species in field boundaries may serve as a 'larder', allowing the beneficial predators to build up in numbers sufficient to keep pest outbreaks in the crop in check. Particularly important predators of cereal aphids include the ground beetles, *Agonum dorsale*, *Bembidion lampros* and *Demetrias atricapillus*, the rove beetles, *Tachyporus chrysomelinus* and *T. hypnorum*, and the earwig, all of which overwinter in field boundaries, especially where a grassy bank is sheltered by a hedge or trees. Other beneficial pest predators that depend to a varying extent on the field boundary for food and shelter, include spiders (especially species of *Erigone* and *Lepthyphantes*), parasitic Hymenoptera (especially species of *Aphidius*), ladybirds (especially the seven spot ladybird) and hoverflies (especially *Episyrphus balteatus*). Alder shelter-belts around orchards encourage the black kneed capsid, a predator of the European red mite, which is a pest of fruit trees.

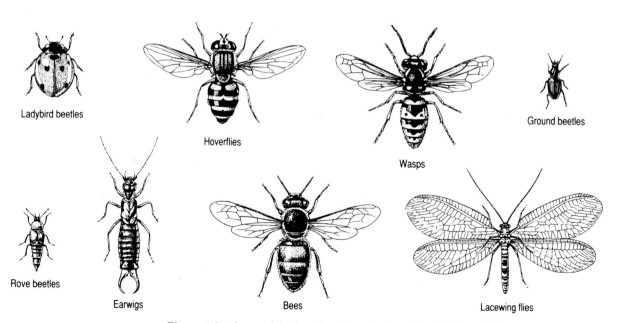

Ladybird beetles

Hoverflies

Wasps

Ground beetles

Rove beetles

Earwigs

Bees

Lacewing flies

Figure 4.2   Some of the beneficial insects found in field boundaries

Some of the larger animal wildlife found in hedges may be viewed as pest species or vermin, but well-managed hedges normally have balanced predator and prey populations. Species such as rooks, crows, pigeons and foxes are highly mobile, so that their occurrence is not determined by the presence of sheltered field margins. Rabbits are fairly easily controlled where they occur along hedges, for example by humane traps, ferreting or shooting. Rat poison should not be used in field margins, because of the damage it would do to populations of species such as shrews and voles, as well as to the animals that prey on rodents.

Disease

For most crops, hedges do not play a significant part in the spread of disease. They have been implicated in the spread of some orchard diseases, since many of the common hedge species, such as hawthorn, blackthorn, crab apple and dog rose, belong to the same family as apple, pear, plum and cherry (the Rosaceae). Fireblight, a bacterial disease which mainly affects pears, is the most economically serious, because infected fruit trees must be grubbed up. The disease is transmitted via the flowers by insect pollinators, so trimming to discourage the flowering of orchard hedges reduces the risk of infection from this source. In any new planting it is possible to avoid inappropriate associations of hedge species and orchard trees.

Diseases, such as potato blight and mildew of wheat and oats, may be encouraged by a build-up of stagnant air in the shelter of hedges. Black rust or stem rust of wheat can be spread from barberry, which is occasionally found as a hedge shrub.

Ditches

Well-managed ditches are of value to agriculture and to wildlife. For agricultural purposes, the priority will be to maintain the drainage function of ditches and, in some areas, their role as a stock barrier, while the conservation objective will be to maintain their wildlife and visual interest. With planning, both agricultural and conservation aims can usually be satisfied. If regular cleaning is needed to allow the free flow of water, it may be possible to carry out the work from one bank only, or on alternate sides for short stretches at a time. Damage to areas of high wildlife value, and the need to remove trees obstructing access, can be avoided in this way, and wildlife will be able to recolonise cleared stretches from the undisturbed areas. Variation in the depth of a ditch over short stretches, to encourage wildlife diversity, is unlikely to affect drainage efficiency.

Trees along the side of a bank, valuable from a landscape and wildlife point of view, may also be useful to the agricultural function of a ditch. The shade provided checks the growth of aquatic vegetation, reducing the need for mechanical cutting or herbicide control. Trees with fibrous root systems, such as willow and alder, help stabilise ditch banks. However, tree roots may damage any nearby underground drains, so trees and drains should be sited apart.

Roadside verges

The main operational considerations in the management of roadside verges are likely to be road safety and maintenance costs. Verges are usually cut to maintain good visibility along the road and prevent the encroachment of scrub and overhanging vegetation. Hedges and trees along the roadside may need to be cut back and any unsafe limbs removed.

The conservation value of a verge can be increased, without prejudicing safety or increasing maintenance costs, by variations in the timing and intensity of the mowing regime to encourage maximum structural and species diversity. Where particularly herb-rich stretches of verge are identified, it may be possible to mark these clearly and single them out for special treatment to maintain and enhance their value.

The use of herbicides and plant growth regulators to control roadside vegetation has now been virtually eliminated, resulting in cost savings for highway authorities and benefits for wildlife.

Railway lines

As with roadside verges, the main commercial considerations in the management of the vegetation along railway lines are likely to be safety and ease of maintenance. Routine maintenance is now kept to a minimum. To keep the track free of vegetation, residual herbicides are sprayed onto the ballast in the summer from a specially adapted train, and a strip alongside the track is occasionally flailed to prevent scrub encroachment and overhanging vegetation which might otherwise obscure signals. Trees may need to be cut back or removed if they are in danger of falling across the track or if there is extensive leaf fall. Accumulations of leaves on the track can result in circuit breaking, trains skidding and the need for expensive repairs to the rolling stock and track.

Where cash limits and safety considerations allow, the incorporation of conservation and landscape objectives into lineside management can bring benefits in the form of more attractive journeys for passengers and improved public relations with those whose property adjoins the line, as well illustrated by the Hampshire Lineside Vegetation Project (Box 4.2).

Income from boundary features

In some cases, boundary features managed for their value as a conservation resource can generate a limited income, or at least result in savings in management costs. They may supply wood products, or an area may be let for shooting. There may also be small, specialist markets for other leisure pursuits such as horse riding, for farm trails, for greenery, such as holly and ivy, used in floristry, and for wild flower and tree seeds. Wide grass verges can be grazed or cut for hay.

Hedgerow trees form a significant proportion of all home-grown hardwood marketed in Britain. The open conditions encourage rapid radial growth, which can result in good-quality, uniform wood, at least in broad-leaved trees, such as ash, beech and sycamore. However, open conditions also encourage side branching, so that unless this has been controlled by regular pruning, only a small proportion of knot-free timber can be obtained from each tree, which will limit their value. For the timber to be marketable, access to the trees must be good. Possible end uses for the wood include veneers, furniture manufacture, construction materials, wood pulp, firewood, fencing and wood crafts. Some research is being done on the use of quick-growing, shelter-belt species, such as willow and poplar, for fuelwood, wood pulp, chipboard, paper and energy production. This has shown that some types of willow can produce up to 40 tonnes of dry matter per hectare every four or five years.

Box 4.2   The Hampshire Lineside Vegetation Project

In the era of steam trains, the vegetation alongside railway lines was intensively managed to reduce the risk of accidental fires caused by sparks from the engines. In summer the banks were scythed and the cut material burned or made into hay. In winter, maintenance gangs carried out hedge laying and coppicing and hand weeded the track. This form of management resulted in the development of fine grassland habitats and well-managed woodland.

The change to diesel and electric trains in the 1960s, shortage of labour and the need to reduce costs led to a much reduced management regime. Apart from the use of residual herbicides to kill vegetation on the track, infrequent mechanical flailing of a narrow strip alongside the line and the felling of obviously unsafe trees, the vegetation was allowed to grow unchecked. The resulting natural succession to scrub and mature woodland now poses a threat to the safety of many lines, and the drastic remedial action needed can meet with public opposition.

To examine the best ways of managing mature lineside vegetation, and to encourage a dialogue on management and conservation issues, the Hampshire Lineside Vegetation Project was set up, jointly funded by British Rail, Hampshire County Council, the Nature Conservancy Council and the Countryside Commission.

In assessing the vegetation, the project gave top priority to the identification of potentially hazardous trees and sites with heavy leaf fall, but conservation and landscape aspects, and the attitudes of local residents, were also considered. The result was a planned and co-ordinated work programme aimed at arresting the vegetation at one of three stages: grassland, scrub or healthy woodland. Tree felling was phased, and new planting with more appropriate species carried out if necessary to maintain a screen between the railway and residential areas. Where appropriate, trees were coppiced or pollarded rather than felled. The flail strip was extended up to 5 metres wide and cut at least once a year to prevent scrub growth and encourage a herb-rich sward. Patches of rare plants were marked so that these could be avoided when spraying and flailing was carried out. In one or two places, trees were felled to re-expose particularly valued distant views, for example that of the local landmark, St Catherine's Hill. Advance planning allowed badger crossings to be incorporated into stretches of track at relatively little cost before electrification of the line took place.

Liaison with residents' groups, local authorities and neighbouring property owners formed an important part of the project. The distribution of information leaflets, public meetings, posters at railway stations and guided visits all helped establish public goodwill.

The introduction of a programme of regular lineside vegetation management, sympathetic to local wishes and conservation needs, can bring a number of significant advantages from the commercial point of view:

- the routine procedures require less organisation and training of staff once they have been set up;

- there is less need for crisis management involving costly contract labour, line closure and compensation if trees fall on neighbouring property;

- with a wider grass strip there is less need for surveillance;

- if trees are thinned while still small there is less need for specialist labour;

- last but not least of the benefits is the more attractive outlook for local residents and railway passengers alike.

The demand for shooting game, for example pheasant and partridge, is greatest for wild birds and in moorland areas with steep valleys. The landowner needs to choose between charging a relatively low rent and leaving all costs to the shooting tenant, or bearing the costs of management in order to achieve a higher rent. Private syndicates may be formed among groups of landowners, with shared facilities and costs. The demand for rough shooting is likely to be highest near large centres of population; the landowner can either make a small charge or allow occasional shooting in exchange for a rabbit or bird control agreement.

Networks of grassy field margins can be exploited for recreational purposes. Organised groups of horse riders, trail bicycle riders, runners and ramblers, and school parties may be prepared to pay a small fee in return for occasional or regular access. Circular routes are most popular, and groups of

landowners can collaborate to provide greater scope and interest. The creation of a **permissive path** need not involve legal procedures and allows the land to be taken back into agricultural use if required, provided that the landowner notifies the Highways Authority that the land is not intended to be a permanently dedicated public right of way. Tracks for riders may need to be of hard-wearing turf, regularly cut and fertilised in order to withstand damage, with riding discouraged in wet conditions. Farm trails will be most successful if they are well signposted and informative. Information can be provided in the form of notice boards, leaflets, school work packs and guided tours by the farmer, farm workers or local naturalists.

There is also a demand for access for leisure pursuits such as cross-country motor-cycle riding and the driving of four-wheeled vehicles, but these are usually incompatible with conservation.

Along transport routes, there is less scope for income generation from conservation management, although attractive routes can be promoted as scenic rides and so generate tourist income for the region. Where verges are wide and there is good access, it may be possible to grow high-quality timber trees as a source of income. Disused railway lines can be developed as nature trails or for use by horse riders. Their generally low gradient can make them appropriate for use by disabled visitors.

Although the income to be expected from boundary features is normally low, it can be increased by the investment of additional management effort. This may be worth while where other enterprises on the farm, such as 'pick-your-own' or livery provision, are complementary, and where grant aid is available for diversification.

4.2   Carrying out your own integrated assessment

The integrated assessment for your own area of land should include the following elements.

▶ A commercial assessment of the area, covering the main points listed in the introductory section of this chapter, and including a map showing the position of boundary features and adjoining land use.

▶ An indication of your attitudes and objectives as they relate to boundary habitat management (or those of the landowner or farm manager if you are an adviser).

▶ Notes highlighting any existing or significant potential interactions between the management of boundary features for commercial purposes and management to maintain or enhance their landscape and habitat value.

The example given in the following section for the case study farm should serve as a guide. With experience, and particularly for small sites, you will probably find it possible to assess the value of an area from both the commercial and the conservation points of view at the same time, and to summarise your integrated assessment with a single, annotated map. Until you are familiar with the assessment process, however, you are less likely to overlook things if you carry out the assessments separately, step by step.

4.3 Commercial and integrated assessment of the case study farm

Commercial assessment

As stated in Chapter 1, the farm covers 368 hectares, with 262 hectares of arable land, 61 hectares of permanent pasture and 16 hectares of woodland. Most of the arable land is in a rotation of two wheats, one barley and a **break crop** of oilseed rape, peas or beans. A suckler herd of 45 red poll cows and a seasonal flock of 120 ewes are farmed on the permanent pasture.

A syndicated shoot rents the shooting rights on the higher parts of the farm on the slopes of Bredon Hill. Gravel is being extracted over a ten-year period from 11.5 hectares at the lower end of the farm. Both these enterprises are kept separate from the farm.

On a day-to-day basis, all agricultural decisions are made by the farm manager. All tractor work and most regular farm maintenance is done by the farm manager and two tractor drivers. Contractors are used to carry out ditching and the shaping of hedges with a circular saw, but both of these are infrequent operations. Management of the woodland and maintenance of buildings is done by two woodmen. Projects with a particular conservation bias, such as the surveying of conservation headlands or the establishment of new hedges or grass margins, are done by a conservation warden employed by a charitable trust, but routine boundary management is part of the farm manager's remit.

The field boundaries do not contribute directly to farm income, and they involve some (unmeasured) costs arising from their maintenance, loss and shading of the crop area, harbouring of weeds, pests and diseases and inconvenience to agricultural machinery. However, they also have agricultural benefits (again unmeasured) by providing shelter for stock, reducing soil erosion, dividing different soil types (although this is now largely historical) and micro-climates, and in the case of tracks, by providing access to fields.

If management decisions were based on purely agricultural grounds, many of the field boundaries on the farm would probably be removed, increasing the field size to at least 20 hectares where possible. However, under the present ownership and management this will not happen, because of the landowner's interest in landscape and wildlife.

Weed ingress into the arable crops from the field boundaries is seen as a significant problem by the farm manager. To combat this, sterile strips are used around most of the arable fields. A width of 1 metre is kept as bare soil by spraying with a herbicide in March–April. A custom-made sprayer has been built to fit onto the front of the farm's mini-tractor for this purpose. Most of the grass and hardcore tracks on the farm are there for agricultural reasons (often along existing footpaths), to provide access to the fields for crop inspection and for heavy machinery, which would otherwise compact the worked soil in the field.

Every year several farm walks are arranged for the public or groups, such as the local Farming and Wildlife Advisory Group or the Royal Society for the Protection of Birds. These do not generate any income, but are seen as a good form of contact with local people and conservation groups.

As elsewhere, the pattern of field boundaries on this farm has arisen to fulfil agricultural functions. However, with the change to more or less continuous arable rotation on some parts of the farm, the agricultural function of the field boundaries has declined. Some hedges have been removed to create more convenient field shapes for large machines.

Despite this, average field sizes on the farm are still small (6 hectares for arable fields and 2.5 hectares for permanent pasture). To overcome the inconvenience of small field size, the arable rotation is farmed in 40-hectare blocks of fields, with gaps in the hedges to allow machinery easy access between fields. There is probably no great economic disadvantage to the small, permanent pasture fields, which would anyway be difficult to amalgamate because of their isolated positions on the edge of the village of Kemerton.

The main impact of agricultural operations on the field boundaries has occurred as a result of physical and chemical disturbance of the hedge bottoms. Past cultivation very close to the hedges, and deliberate use of herbicides to control hedgerow weeds, have resulted in the replacement of the perennial flora with problem annual weeds, such as cleavers and barren brome, which have little value for wildlife. This has led to the use of the sterile strips already mentioned. Protection of the hedge bottoms and track sides from disturbance is therefore a high priority in agricultural as well as conservation terms. Once disrupted, it is difficult to return the field edge flora to its previously undisturbed state, despite careful management.

Management of ditches and streams is relatively small scale and infrequent apart from the trimming of bank sides in autumn. This has a conservation benefit by favouring the finer-leaved grasses and wild flowers over the coarser vegetation that would otherwise block the ditches.

Objectives and constraints

The broad objectives of field boundary management on the farm are:

▶ to maintain the existing boundary pattern, paying special attention to the protection and improvement of the boundaries that contain diverse or rare plants and animals, that are visually prominent or that connect other important habitats and features;

▶ to keep boundaries around pasture stockproof;

▶ to minimise weed ingression into the crop;

▶ to keep boundaries looking well managed.

An example of one of the ways in which objectives and constraints can be examined in more detail is shown in Figure 4.3, which is based on an approach described in Chapter 5 of the foundation book. Figure 4.3 shows the factors relating to the management of boundary features as perceived mainly by the farm manager, but in consultation with the landowner. Factors outside the circle are those that the farm manager considers to be outside his control, which therefore constitute possible constraints. Those inside the circle are under his control or could be if he took appropriate action. These relate to his objectives. Arrows indicate factors that are changing: moving into or out of control. The figure emphasises the importance that the farm manager places on a well-kept appearance to the farm, including weed control for both cosmetic and agricultural reasons. This may restrict some of the conservation options. The use of an exercise such as this allows open recognition and discussion of any differences in objectives and perceptions among those involved. It thus provides an opportunity for any conflicts to be resolved, so that the resulting management plan is more likely to be successfully implemented.

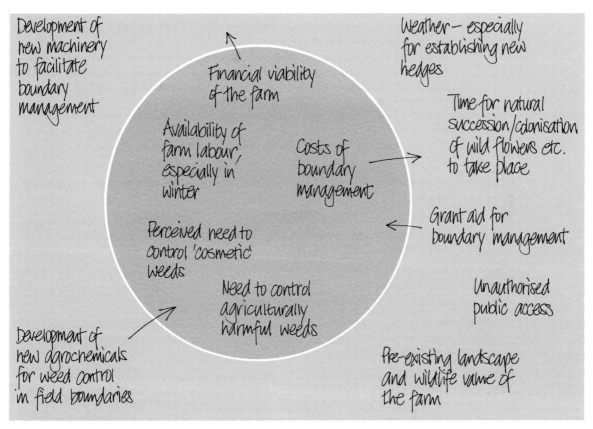

Figure 4.3   Factors affecting field boundary management on the case study farm

Chapter 5
MAINTAINING AND IMPROVING EXISTING BOUNDARIES

Once you have made a comprehensive assessment of the boundary features on your land, you will have completed Stage 1 of the management planning cycle shown in Figure 1.2, and will have a good basis for deciding the best options for the future management of the features. Stage 2 involves identifying clearly your objectives for the site, and any constraints that apply, as referred to briefly in the case study example at the end of Chapter 4 and described more fully in the foundation book. Chapters 5 and 6 are concerned with Stage 3 – identifying the range of options available to help you to achieve your objectives and selecting the most appropriate ones for your site.

Appropriate conservation management options for boundaries fall into three categories:

▶ *maintaining* a boundary in its present form because it is a valued landscape feature, because it contains rare plants and/or because it is old and has a high ecological value;

▶ *improving* a boundary of moderate or poor conservation value by changing its visual appearance, increasing its structural and species diversity and/or encouraging particular wildlife species;

▶ *creating* a new boundary feature to add landscape interest and/or support and encourage a wide diversity of wildlife or particular wildlife species.

If you have time, it is worth investigating as many management options as possible before narrowing the choice down to those that correspond most closely to your own objectives and constraints. You can add to your own ideas by reading trade magazines, noting what other landowners have done, seeking the advice of experts (see the *Helpful Organisations* supplementary booklet in the foundation module) and talking to friends and acquaintances (even those who you suspect may have ideas at odds with your own). For each boundary type, such as a hedge, bank or verge, the options are likely to be similar whether the boundary occurs on farmland or along a transport route. Only the constraints governing the choice of options will differ. In all situations, options that minimise any damaging disturbance to a boundary from adjoining land use will have conservation benefits.

Chapter 5 describes the principal options available for maintaining and improving boundaries that already exist, while ideas for creating new features are given in Chapter 6. Inevitably, there is some overlap between the two, although with new features there is more opportunity to select for specific objectives. The process of exploring and choosing options for both existing and new features on the case study farm is described at the end of Chapter 6.

5.1   Hedge management options

Hedges can be managed by trimming, laying and coppicing, but there is considerable variation in the way in which these three basic options can be implemented and combined. Different forms of management benefit different groups of wildlife and add variety to the landscape, but the following general guidelines should be kept in mind:

▶   avoid trimming hedges when birds are nesting (mid-March to mid-July, slightly later in the north);

▶   ensure a good supply of fruits and berries as a source of food for wildlife in winter by trimming hedges less frequently than once a year, or as late in the winter as possible;

▶   avoid damaging the perennial vegetation at the base of hedges by protecting it from pesticide and fertiliser drift, cultivation, over-grazing and burning;

▶   encourage diversity in hedge structure by, for example, allowing some hedgerow saplings to grow into trees, or varying the management of different sections of hedge;

▶   plan hedge management on a rotational cycle so that there are always stretches of hedge left undisturbed, which can act as a refuge for wildlife and a source of recolonisation;

▶   where possible, adopt the local traditional form of hedge management if this is an important regional landscape characteristic.

Trimming

Variations in the frequency and timing of trimming, the shape and size to which a hedge is cut and the removal or retention of trimmings, dead wood, saplings and mature trees will affect the conservation value of a hedge.

Frequency

Since most common hedgerow species flower and fruit more freely on two- or three-year-old wood, hedges will be most valuable as a source of winter food for wildlife if they are trimmed less frequently than once a year. Where annual trimming is desirable for any reason, delaying the operation until late in the winter will allow wildlife to exploit the food source first. Trimming every two or three years should be sufficient to maintain a dense, stockproof structure, while not allowing the wood to thicken to the stage where it is difficult to trim. A rotation should be established, so that only some of the hedges, or alternate sides, are trimmed in any one year, to minimise the disturbance to wildlife (as well as spreading the workload).

It is possible to maintain hedges without any regular trimming, by coppicing or laying them once every ten years or so and allowing them to grow freely in the intervening years (see below).

In some cases, frequency of trimming is governed by specific commercial constraints, as mentioned in Chapter 4. For example, annual trimming may be needed for roadside hedges to maintain safety and for some orchard hedges to discourage flowering and so reduce the risk of fireblight infection. Frequent trimming may be needed to comply with the conditions of a tenancy agreement, but if this is the case it may be possible to renegotiate the agreement to meet conservation aims.

From the conservation point of view, trimming is best left until late in the winter (December–February), by which time most of the fruits and berries will have been eaten. However, it should not be undertaken in frosty weather, since this may damage the hedge. The bird nesting season, from about mid-March to mid-July, should always be avoided. There may be special characteristics of the hedge shrubs that dictate when trimming occurs. For example, beech and hornbeam hedges retain their dead leaves longer, and so can provide more shelter from wind, if they are cut in late summer. Holly, being evergreen and frost sensitive, is best clipped in August or September before there is any risk of frost damage.

Size

In general, the greater the volume of a hedge, the more food and shelter it will provide for wildlife, so the conservation value of a small hedge can be improved by reducing the intensity or frequency of trimming. A height of 1.5–2 metres and a width at the base of 2.5 metres will satisfy the requirements of a great variety of wildlife. However, if your aim is to encourage a specific group of wildlife, you may need first to determine the particular requirements of the group and use this as the basis for determining hedge size. If you want to encourage ground-nesting gamebirds, such as pheasant and partridge, keep hedges relatively narrow and low (below 2 metres) to avoid shading out the grassy vegetation at the base. Tall hedges with long outgrowths and hedgerow trees may provide lookout posts for egg predators, such as the magpie, and so will be avoided by partridges seeking well-protected nesting sites.

Shape

Hedges may be trimmed to a variety of shapes: 'A'-shaped, rectangular, chamfered, topped 'A' or rounded (Figure 5.1). The 'A' shape is most often recommended for conservation purposes, since the gradation in width provides structural diversity at successive levels of a hedge and the wide base provides good cover for many forms of wildlife. Other advantages are that the 'A' shape, with its narrow top, reduces self-shading, minimises wind turbulence, prevents any damaging accumulations of snow and makes it easy to avoid cutting saplings singled out as potential hedgerow trees. If for any reason it is necessary to restrict the width of a hedge, an asymmetric 'A' shape could be considered, with, if possible, the wider base on the sunnier side of the hedge.

Where space is restricted, for example where a hedge is on a bank or alongside a ditch, or where there is a rich and attractive ground flora at the base, or the aim is to encourage ground-nesting birds, a rectangular-shaped hedge with a narrower base may be required. Those who favour a rectangular shape claim that hedge growth is promoted by the flat top

Rectangular   'A'-shaped   Chamfered   Topped 'A'   Rounded

Figure 5.1   Hedge shapes

exposed to the sun, but disadvantages are that growth lower down may be checked by shade so that gaps develop, and trimmings and snow may accumulate on the flat surface and weaken the structure of the hedge. Rectangular-shaped hedges may need to be laid periodically if they are to remain stockproof at the base.

The chamfered and topped 'A' shapes result in a bulky hedge, which is good as a stockproof barrier and as shelter for wildlife, but may require more passes of the cutter than the simple 'A' shape. Rounded hedges are usually the result of hand trimming, and so normally confined to small stretches of hedge near houses, but are the characteristic local style in certain areas of Wales such as Powys. Whatever the hedge shape chosen, each successive trimming should be positioned slightly beyond the previous cut to encourage a densely branched structure (see Figure 5.2). Cutting back to the same place each time will remove many of the new branching points, and the tips of the old shoots become progressively more woody and lose vigour. Every so often a more severe cut may be needed to keep a hedge to the required size. It is better to do this with a shape saw rather than a flail.

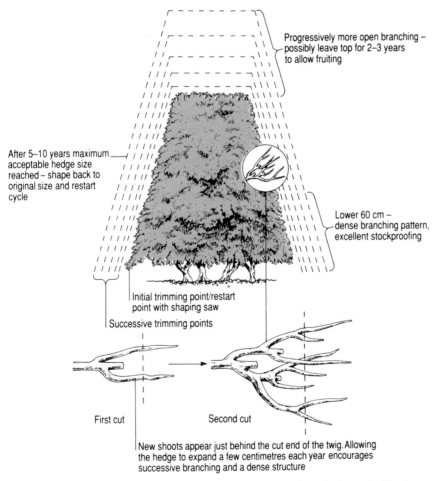

Figure 5.2   Pattern of trimming required to develop a densely branched hedge. (Source: Shropshire Farming and Wildlife Advisory Group leaflet)

67

Additional measures to encourage wildlife, which can be taken at the time that a hedge is trimmed, include retaining some of the trimmings and dead wood, and singling out saplings or some of the shrubs which make up the hedge to allow them to develop into hedgerow trees.

Even when trimming is done in mid-winter to minimise the disturbance to wildlife, some groups of wildlife will be adversely affected, for example the brown hairstreak butterfly, whose eggs overwinter on blackthorn. If some of the larger cut material can be left stacked in piles out of the way of paths but near the hedge, this provides a chance for some of the affected insects to recolonise the hedge later, as well as providing additional habitat. Dead hedgerow trees, such as elm, which may be considered an eyesore and so removed when hedges are trimmed, are of considerable value for wildlife such as insects, spiders, fungi, woodpeckers and owls, so consider retaining them if they are not a hazard and have little marketable value.

Laying

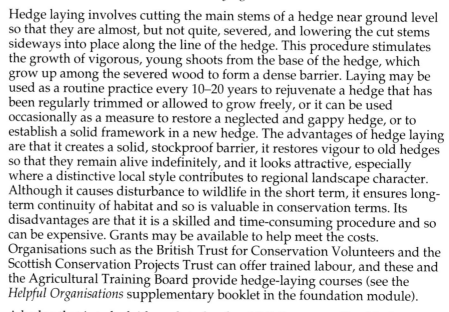

Hedge laying involves cutting the main stems of a hedge near ground level so that they are almost, but not quite, severed, and lowering the cut stems sideways into place along the line of the hedge. This procedure stimulates the growth of vigorous, young shoots from the base of the hedge, which grow up among the severed wood to form a dense barrier. Laying may be used as a routine practice every 10–20 years to rejuvenate a hedge that has been regularly trimmed or allowed to grow freely, or it can be used occasionally as a measure to restore a neglected and gappy hedge, or to establish a solid framework in a new hedge. The advantages of hedge laying are that it creates a solid, stockproof barrier, it restores vigour to old hedges so that they remain alive indefinitely, and it looks attractive, especially where a distinctive local style contributes to regional landscape character. Although it causes disturbance to wildlife in the short term, it ensures long-term continuity of habitat and so is valuable in conservation terms. Its disadvantages are that it is a skilled and time-consuming procedure and so can be expensive. Grants may be available to help meet the costs. Organisations such as the British Trust for Conservation Volunteers and the Scottish Conservation Projects Trust can offer trained labour, and these and the Agricultural Training Board provide hedge-laying courses (see the *Helpful Organisations* supplementary booklet in the foundation module).

A hedge that is to be laid needs to be about 2.5–5 metres tall, with the main stems ideally 5–10 centimetres in diameter at the base. Before a trimmed hedge can be laid, it needs to be allowed to grow up for between two and five years. The stages in the process are illustrated in Box 5.1. Operations begin by clearing shoots and wood from outside the hedge line and removing any dead wood, old fencing or wire. Climbers that may smother the hedge, such as old man's beard, hedge bindweed, black bryony and white bryony, and elder, which is invasive, are usually dug out or at least cut back, although briars may be kept. Clean, straight, pliable but woody stems with twiggy tops are selected and their lower stems cleaned of side shoots. If a hedge is sparse, every available stem is used, but otherwise unwanted stems are removed from around the selected shoots (although a few may be left in reserve in case the selected stems split). The selected stems are partially severed near ground level. These cut stems, or **pleachers**, are then laid sideways in one direction at an angle of 30–40 degrees, and the branches usually secured with stakes at intervals to strengthen the structure.

Box 5.1   Laying a hedge (Midlands style)

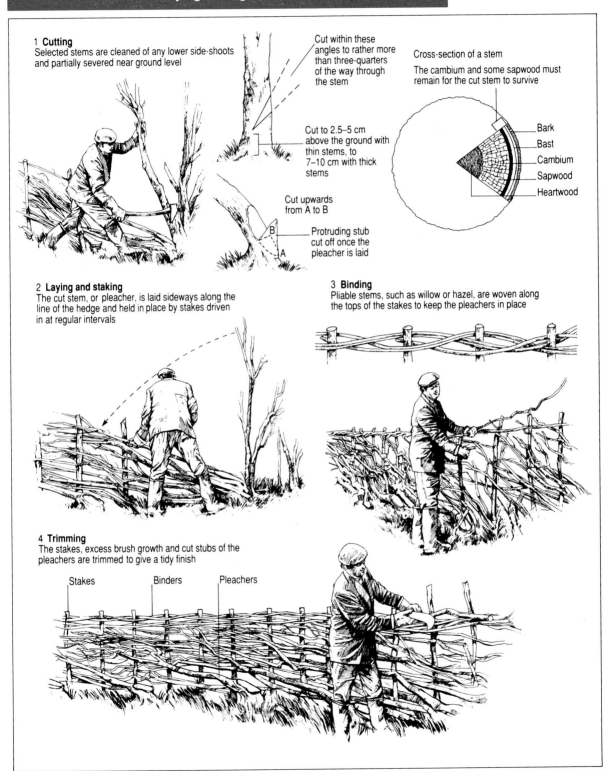

1 Cutting
Selected stems are cleaned of any lower side-shoots and partially severed near ground level

Cut within these angles to rather more than three-quarters of the way through the stem

Cut to 2.5–5 cm above the ground with thin stems, to 7–10 cm with thick stems

Cut upwards from A to B

B

A

Protruding stub cut off once the pleacher is laid

Cross-section of a stem

The cambium and some sapwood must remain for the cut stem to survive

Bark
Bast
Cambium
Sapwood
Heartwood

2 Laying and staking
The cut stem, or pleacher, is laid sideways along the line of the hedge and held in place by stakes driven in at regular intervals

3 Binding
Pliable stems, such as willow or hazel, are woven along the tops of the stakes to keep the pleachers in place

4 Trimming
The stakes, excess brush growth and cut stubs of the pleachers are trimmed to give a tidy finish

Stakes   Binders   Pleachers

Pliable stems, termed **edders**, ethers, binders, heathers or heatherings, such as those of willow or hazel, are usually woven along the top to prevent the branches springing out of place or being pushed up by livestock. After a branch is laid, the exposed stub, where the cut was made, is trimmed to remove any snags, and the stakes and any excess brushy growth cut back if necessary to the required height and width.

Laid hedges are most commonly found in the Midlands, Wales and parts of the South-West. There is considerable regional variation in style, depending on the purpose of the hedge and the individual hedge layer. In 'bullock fencing', designed to keep cattle in, which is the predominant style in the Midlands, the laid hedges are 'single brushed', with all the pleachers laid to one side and the brushy tops towards the cattle, so that the young shoots are protected from grazing. Many Welsh hedges are 'double brushed', with the pleachers laid to both sides, producing a bushy hedge to keep in sheep. Some Welsh hedges are thin and single brushed, but reinforced with dead wood cuttings, whereas in other areas hedge layers frown on the use of dead wood. In Wiltshire and Dorset, no stakes or edders are used, and the pleachers are spread out from the central line to create a dense, wide hedge. Detailed descriptions of regional variations are given in *Hedging* (British Trust for Conservation Volunteers, 1975).

Not only hawthorn, but most types of hedgerow shrub, for example elm, blackthorn, ash, maple, sycamore, hazel and willow, can be laid. Holly can also be laid, but it is brittle and can easily be damaged if laid in frosty weather; it should only be laid in May.

Filling gaps

Gaps in a hedge where dry grass and nettles occur provide good nesting sites for pheasant and partridge, so may be left as they are on arable farms where there is an interest in game. Relatively small gaps in a hedge can be bridged by pleachers during the hedge-laying process. A more permanent barrier can be achieved by **layering,** or encouraging a pleacher to take root in a gap by pegging it down so that it is in contact with the soil (Figure 5.3).

Cut the pleacher so that it lies close to the ground with some part of it touching the earth. Clear away any grass at the point of contact and dig out a shallow trough, so that the stem is half buried in the earth just at this point

Alternatively, cut a notch in the top of the pleacher where it lies close to the ground and heap earth over the notched section, burying it

Figure 5.3
Layering to fill a small gap in a hedge. (Source: British Trust for Conservation Volunteers, 1975)

If the weight of the pleacher is not enough to hold it in position, peg it down so that it is in firm contact with the ground

Roots should develop where the stem is buried. When the hedge is next laid, the stem can be cut through between these roots and the old ones, creating a new and independent plant

Large gaps will need to be filled by new planting. This is most likely to succeed if shade-tolerant species, such as holly, which are able to compete with the existing shrubs, are used. Blackthorn is also useful for filling gaps in a hawthorn hedge. If hawthorn is to be used to repair an existing hawthorn hedge, it should be planted into fresh soil. Cutting back or coppicing a hedge either side of a gap, digging in well-rotted manure and using larger **transplants** than for a new hedge will all increase the chances of survival of replacement plants.

Coppicing

On arable farms, where a stockproof barrier is not required, or in cases where the wood is not pliable enough for laying, a neglected or overgrown hedge can be managed by coppicing. This involves cutting the hedge down to just above ground level (to about 7.5 centimetres), and can be done either with a hand-held chain saw or with a tractor-mounted circular saw. The cut should be horizontal or, in the case of multiple stems or a single very large stem, angled up towards the middle, to prevent water collecting in the centre and causing the stump to rot (Figure 5.4). New shoots should spring up from the old base, although up to 5% of stumps may die unless the operation is carried out very carefully. Species that respond particularly well to this form of management are hawthorn, hazel, hornbeam, lime, oak and sweet chestnut. Species such as alder, beech, wild cherry and poplar only coppice well when young. Holly and blackthorn do not respond well.

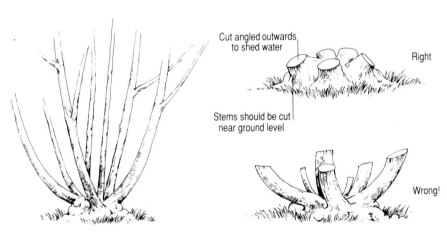

Figure 5.4   Coppicing

The advantages of coppicing are that it requires less skill and is less time consuming than laying, so it is a relatively low-cost management option. In parts of the country, it may once have been the routine form of hedge management, used to provide regular supplies of wood products, such as firewood and stakes.

Coppicing has a dramatic, if temporary, effect on the conservation value of a hedge, in terms of both landscape and wildlife, so that only short stretches of hedge should be tackled in any one year. Where a hedge is highly visible, it may be worth publicising the purpose of this form of management, perhaps with a temporary notice next to the hedge, or a note in the local paper or parish magazine, so that people understand why it is being done.

Very old hedges, with a rich mixture of shrub species, may need to be managed by a combination of laying, coppicing and trimming, with some species left as standard trees, depending on the component shrubs and their condition.

Hedgerow trees

Tagging saplings

The presence of hedgerow trees adds to the conservation value of a hedge by increasing its structural diversity. Where appropriate, new hedgerow trees can be established in existing hedges by selecting sturdy, straight-stemmed saplings in the centre of the hedges and marking them clearly with plastic tags (strips of old fertiliser bag will do), so that the hedge cutter can avoid them. Selection is best done in the summer when the saplings can be seen in leaf. Several shoots can be left at each point, and the strongest selected when the hedges are trimmed. In a hedge which has previously been laid, regrowth can be selected, but the shoots chosen should lie above the original base to avoid weak, 'S'-shaped stems. It is essential to ensure that the selected shoots are well away from any overhead cables.

Notch planting

If no suitable saplings or sturdy stems occur where a tree is needed, new trees can be established by planting. Since competition for light and nutrients from the existing vegetation makes tree establishment difficult, natural gaps in a hedge should be exploited or a notch cut into the side of the hedge, as shown in Figure 5.5.

Transplants generally survive better and grow faster than larger plants, or **whips**. Good weed control will be needed round the base of the young tree for

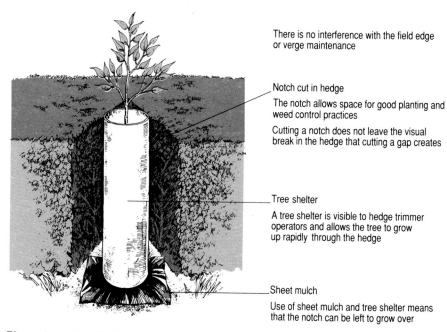

There is no interference with the field edge or verge maintenance

Notch cut in hedge
The notch allows space for good planting and weed control practices
Cutting a notch does not leave the visual break in the hedge that cutting a gap creates

Tree shelter
A tree shelter is visible to hedge trimmer operators and allows the tree to grow up rapidly through the hedge

Sheet mulch
Use of sheet mulch and tree shelter means that the notch can be left to grow over

Figure 5.5   Establishing a new hedgerow tree by notch planting. (Source: Forestry Commission, 1990)

the first few seasons, and can be achieved by planting into a plastic **mulch** or by careful herbicide use. Tree shelters, if possible reaching above the height of the hedge, will encourage a straight, well-formed tree, protect it from browsing animals and help ensure that it is avoided when the hedge is trimmed.

Oak, ash, wild cherry, field maple, hornbeam and holly all make attractive hedgerow trees and are good for wildlife. Birch and poplar are fairly short lived. Beech, sycamore, horse chestnut and conifers cast dense shade. To ensure that the number of saplings established will be sufficient to provide continuity in the number of mature hedgerow trees, a ratio of six saplings, three small trees and two medium-sized trees is needed for each mature tree.

Routine maintenance

Hedgerow trees should be inspected from time to time to ensure that branches that might pose a hazard to people, vehicles or machinery are removed. If the trees are intended for timber, branches below a height of 4 or 5 metres (2.5 metres is acceptable for veneer-quality oak and sycamore) should be removed at an early stage so that the wood is free of knots. The trees should be protected from damage by vehicles and machinery, and should not be used as living fence posts.

Harvesting mature trees

Timber merchants may be reluctant to buy timber from hedgerow trees because of the risk of damage to their expensive machinery from metal objects, such as fencing wire and nails, embedded in the wood. Many of the other considerations are the same as in the harvesting of timber from woodlands (see *Woodlands*). In particular, access for the removal of the timber must be adequate. Legal constraints may apply. A Forestry Commission licence is needed to fell more than 5 cubic metres of timber, or to sell more than 2 cubic metres, in any three-month period. Occasionally hedgerow trees that are important local landmarks are protected from felling by a Tree Preservation Order. On tenanted farms, all timber, trees and saplings belong to the landlord. Windfall timber also belongs to the landlord, although in some districts, such as Norfolk, the tenant has the right to loppings. Decayed or dead trees when blown down belong to the tenant.

Pollarding

Some hedgerow trees, and riverside willows, have traditionally been managed by pollarding. In this form of management, a tree is cut 2–3 metres above the ground, producing new shoots out of reach of browsing stock. Pollarding can be used to provide a regular source of firewood and, in the case of willow, **rods** or **withies** used in basket work. Many oak and ash pollards mark old parish or district boundaries. Old pollarded trees have great conservation value, not only because of their historic and landscape significance, but because pollarding prolongs the life of a tree indefinitely, providing a stable environment for a rich variety of wildlife, such as insects, birds, lichens and ferns.

Willow can be pollarded either annually to provide rods or withies, or less frequently for poles or firewood. For wildlife purposes, a long interval between successive pollardings is preferable to minimise disturbance. An interval of 20–40 years provides a balance between maintaining the vigour of a tree and stability for its wildlife community.

Young trees can be pollarded by cutting the main stem horizontally at the required height once the stem is 10–15 (but not more than 30) centimetres in diameter. Old pollarded trees can be re-pollarded by cutting the stem immediately above the **callus**, or wound tissue, of the previous cut and cleaning out debris from the crown to discourage rotting.

5.2 Stone walls

Well-built stone walls should require relatively little maintenance unless they are damaged by stock, burrowing animals, vibration or direct impact by traffic, accidental or deliberate damage by people, exceptionally severe weather or air pollution. Regular inspection to ensure that any loose stones are firmly wedged back in place will help limit damage, and may be all the maintenance required. Over-enthusiastic re-pointing of mortared walls may prevent plants becoming established, by smoothing over the cracks where they might otherwise gain a hold. Although woody vegetation can damage the structure of a wall and may need to be removed, other plant growth does little harm and adds to the conservation value of the wall.

Repairing gaps

The basic procedure for building or repairing a dry stone wall is described in Chapter 6, Box 6.1, but it is a skilled task and except for small repairs is likely to require expert help. Training is provided by the Dry Stone Walling Association of Great Britain and the Agricultural Training Board as well as by the British Trust for Conservation Volunteers and the Scottish Conservation Projects Trust (see the *Helpful Organisations* supplementary booklet in the foundation module). Where a small section of wall has collapsed, the damaged section can be dismantled until a firm section on either side is reached, and the wall rebuilt by copying the existing pattern. The stones near the base may be undamaged, in which case they can be left in place. The gap should be stepped, so that the new section can be firmly meshed in with the old. In the case of earth-filled stone walls, it may only be necessary to rebuild one side or, if both sides are damaged but the centre remains firm, one side at a time can be rebuilt. Replacement stones should be from local sources wherever possible, recovered from the collapsed section, from derelict walls or buildings nearby (provided these are not themselves of conservation value) or from a local quarry. Any moss or other vegetation on old stones should be retained unless its removal is essential to ensure a good fit; it can be at least three years before lichens and ferns start to colonise a fresh stone surface.

If a wall leans or bulges, this may be because the **foundation stones** are badly positioned or have moved. In this case, the affected section needs to be dismantled to the base and the foundation stones repositioned before the wall is rebuilt.

If your wildlife assessment reveals that any particularly interesting or unusual plants have colonised small, tumbledown sections of wall, it may be preferable in conservation terms to leave the wall as it is rather than rebuild it.

In summary, the main points to keep in mind for conservation purposes in the care of stone walls are:

▶ check walls regularly for loose stones so that they remain in a good state of repair;

▶ use local sources of stone for repairs wherever possible;

▶ retain any moss or other vegetation on old stones unless it affects the stability of the wall;

▶ consider leaving small sections of collapsed wall as they are if they have been colonised by unusual plants;

▶ when re-pointing mortared walls, leave some crevices as habitat for animals and plants.

5.3   Ditches

Ditches which have a variety of wildlife and which continue to fulfil their drainage function may need little management attention, apart from protecting them from the polluting effects of pesticide and fertiliser drift, and silage and slurry effluent. Maintaining a wide strip of bankside vegetation provides a buffer zone that helps avoid contamination from adjoining land use, and prevents the soil slippage that can occur if cultivation is carried right up to the edge of a ditch.

Once every five to seven years, depending on the nature of the substrate and the flow rate, it may be necessary to dredge or re-profile water courses to remove accumulations of silt and vegetation if they are impeding water flow. In some cases, it may be necessary to control excessive weed growth more frequently.

Dredging

Although necessary both to the function of a ditch and for the maintenance of this habitat type, dredging needs to be carried out carefully if it is not to result in permanent damage to the wildlife interest. There are a number of measures that can be taken to ensure that the disturbance to wildlife is minimised and opportunities for recolonisation provided. As with hedge cutting, dredging is best planned on a rotational basis, so that only a proportion of the total length of ditches on a holding is tackled in any one year. Working from one bank only leaves reservoirs of undisturbed habitat. Autumn cleaning is preferable to summer, if time and weather conditions permit, or the work can be staggered through the year, avoiding the months from March to July to prevent disturbance to breeding birds.

Where the vegetation is particularly interesting, consider leaving short stretches untouched, or keep the spoil with the plants to one side and replace them once the work is done. Leave some patches of weed as nesting areas. Where the ditch side has recently become dominated by aggressive weeds, such as cleavers and stinging nettles, it may be possible to encourage the return of a more diverse flora by removing some of the enriched topsoil with the vegetation. The establishment of a buffer zone alongside the ditch will help prevent further enrichment from fertiliser drift and run-off.

The silt that is removed as a result of dredging is usually high in nutrients, and so should not be dumped on the banks where there is interesting bankside vegetation. Nor should it be put in bankside hollows used by breeding waders and wildfowl. Ideally, it should be spread on adjoining fields, where the increase in fertility should be of benefit and any aggressive weeds can be controlled during normal field operations. Alternatively, it can be spread thinly along the less interesting areas of bankside.

Ditches where the banks are gently sloping provide conditions suitable for a variety of wildlife. Dredging provides an opportunity to alter the profile of steep-sided ditches to encourage greater wildlife diversity (Figure 5.6). Where there is space, both banks can be pulled back and graded to reduce the slope. If space is restricted, for example if there is a hedge alongside the ditch, only the side away from the restriction need have a shallow slope or, if there is a choice, only the sunnier side. If a ditch is wide, it may be possible to increase structural diversity by dredging only one half, leaving a muddy ledge or shallow underwater **berm** along its length (Figure 5.6). This option is particularly useful where the conservation value of a ditch is high, since it

75

can be used as a way of leaving stretches undisturbed, thereby increasing opportunities for recolonisation in cleared sections.

Where possible, some variation in the profile of water courses should be maintained. Sections of bank with a vertical face provide nest sites for kingfisher, sandmartin and water vole. Areas where the bank has slipped provide muddy areas that are especially valuable for invertebrates. Even margins trampled by cattle have their own value; the wet, disturbed ground can provide a home for specialised species, such as the nationally rare adders-tongue spearwort. Variations in water depth encourage a diverse aquatic wildlife and need not impede drainage function over short sections. Areas of deep water can provide a safe haven if the surface water freezes or if the main channel dries out. Variation in the depth of a water course can be introduced over short stretches when dredging work is being carried out (Figure 5.7).

Vegetation control

Where control of aquatic vegetation is necessary, hand or mechanical cutting is usually preferable to the use of chemicals. If chemical control is essential, only herbicides approved for use in water courses should be used, and expert advice should be sought, for example from the National Rivers Authority (see the *Helpful Organisations* supplementary booklet in the foundation module). In some cases, careful and limited use of approved chemicals can cause less disturbance to wildlife than mechanical cutting.

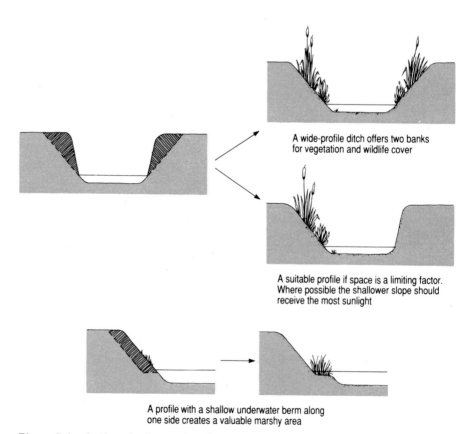

A wide-profile ditch offers two banks for vegetation and wildlife cover

A suitable profile if space is a limiting factor. Where possible the shallower slope should receive the most sunlight

A profile with a shallow underwater berm along one side creates a valuable marshy area

Figure 5.6   Options for increasing the conservation value of a ditch by changing the profile

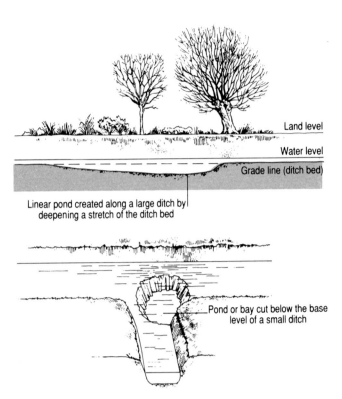

Land level

Water level

Grade line (ditch bed)

Linear pond created along a large ditch by
deepening a stretch of the ditch bed

Pond or bay cut below the base
level of a small ditch

Figure 5.7   The creation of a deep section (or linear pond) when dredging a ditch

Cutting is most effective as a method of weed control if it is carried out in
June or July when growth is most vigorous, although autumn cutting is less
disturbing from a conservation point of view. Where frequent cutting is
necessary, it is best done on a rotational basis, or alternate sides tackled in
turn. Whether weed is cut or chemically controlled, the treated vegetation
should be removed from the water, since it may otherwise impede drainage
and deplete the water of oxygen as it decays.

Wherever possible, care should be taken to avoid damaging patches of rarer
and less vigorously growing species. Examples include water plantain,
flowering rush, water violet, yellow flag, water-lilies, river water-dropwort,
broad-leaved pondweed, giant water dock, arrowhead, club-rush,
unbranched bur-reed and reedmaces.

Bankside trees can help shade out excessive weed growth. For this purpose,
a ratio of about 6:4 between shaded and unshaded water is recommended to
control vegetation without discouraging all growth. Occasional thinning of
bankside trees may be needed to encourage the development of those that
remain. Branches and pieces of timber that collect in the water can serve as
nesting sites for moorhen and coot, so need only be removed if they are
impeding the water flow.

Bankside trees

To summarise the conservation guidelines for ditch management:

▶   avoid contamination by pollutants, such as pesticide, fertiliser drift and
seepage, slurry and silage effluent, oil, rubbish or excessive rotting vegetation;

▶ plan weed cutting and dredging on a rotational basis, so that a proportion of the water course is undisturbed;

▶ carry out work from one bank only, or alternate banks along different stretches;

▶ spread excavated spoil thinly, away from areas of high conservation value;

▶ maintain and enhance structural diversity where possible, e.g. retain or create cliffs and muddy ledges;

▶ avoid non-selective chemical weed control.

5.4   Banks and verges

Management options for the maintenance of banks and verges relate mainly to the mowing regime and scrub control. In many cases, hedge, ditch and tree management will also be involved, but the options are likely to be similar to those already discussed.

If a bank or verge is of high conservation value, the aim should be to continue or imitate the form of management that has led to its interest. For example, flower-rich banks with fine grasses need mowing at least annually to retain their character. If the area is of moderate or low conservation value, varying the cutting regime across its width will increase the plant and structural diversity and so enhance its value.

As with all boundary habitats, it is important to minimise any damaging effects from adjoining land use. Drift from pesticides and fertilisers should be prevented, and any temporary dumping of materials, such as grit for roads and crop debris, should be restricted to areas of little conservation value.

Mowing regime

The timing and intensity of the mowing regime will have an important influence on the composition of the bank and verge flora, and so for conservation purposes should be closely related to the assessment of the wildlife value of a site.

Infrequent cutting, on a two- or three-year rotational cycle, will minimise the disturbance to wildlife, encourage perennial grasses and flowers and allow dead grass to accumulate, providing shelter for insects, small mammals and ground-nesting birds, such as partridge. Where annual mowing is preferred, to control tall vegetation and scrub and prevent the accumulation of dead grass, this is best done in late autumn to minimise disturbance and allow most plants to set seed. On roadsides where an unobstructed view of roadsigns is important, and at dangerous bends and junctions, vegetation may need to be mown at least twice a year, once in May or June when growth is most vigorous and again late in the autumn. If frequent mowing is necessary in flower-rich areas, setting the cutter bar to about 30 centimetres will allow smaller plants to flower and set seed. More regular mowing may be necessary where a verge is used as a footpath, or where access is required for vehicles and machinery.

Where a bank or verge has a high conservation value, modifications to the mowing regime can help ensure that the site retains its interest. Often this involves mimicking traditional management practices that allowed certain plants to set seed while discouraging seed formation in more vigorous competitors. For example, some sites are valuable because they have a flora typical of grazed chalk grassland: fine grasses and flowers such as cowslip, restharrow and birds-foot-trefoil. These flourish with fairly regular cutting to restrict coarser vegetation, providing that the cutter bar is set high enough to allow them to seed. Where the flowers are typical of those found in a hay meadow, mowing at the time that a hay crop would normally be taken helps maintain the characteristic composition. If the aim is to encourage a rich variety of fine-leaved and poorly competitive plants, raking up and removing cuttings and dead vegetation will prevent these being smothered.

Where the existing conservation interest of banks and verges is moderate, it can be enhanced by varying the mowing regime on different parts of the site or, where there is space, across the width. For example, a roadside verge might be mown in the following way: twice annually, or more frequently if necessary, in a strip along the road edge as a footpath and to improve visibility for drivers; once a year in the middle section after most plants have set seed to encourage a variety of flowering plants; and only occasionally in the area furthest away from the road to allow tall vegetation and accumulations of dry grass while preventing the encroachment of scrub.

Tree management

In many cases, tree management options on banks and verges will be similar to those already discussed for hedgerow and bankside trees, but where verges are wide, as along motorways and some railway lines, denser stands of trees may occur. From time to time, thinning may be needed to allow the remaining trees space to grow strongly and to remove those that are unsafe. Provided that they do not represent a hazard, fallen trees can be left where they are. Where trees are important as a screen, a programme of selective felling and replacement of mature and over-mature trees will be needed to ensure continuity. For some species, such as hazel, phased coppicing of small areas at a time on a long rotation will help maintain tree cover.

Where only one or two tree species have been used in previous planting schemes, other, native trees and shrubs can be interplanted to increase the wildlife diversity and add visual interest. Guidance on the selection of species is provided in Chapter 6. Where access is good, it may be worth considering the use of species from which a marketable crop can eventually be obtained.

5.5 Deciding on options for managing your own boundaries

The exercise and case study example that relate to this chapter are included with those at the end of Chapter 6.

CREATING NEW BOUNDARY FEATURES

Many existing boundary features of high conservation value, such as old parish boundary hedges, owe their diverse and interesting flora to gradual colonisation over very many years, or to the fact that they were established at a time when the variety of nearby seed sources was far greater than it is nowadays. For this reason, if limited resources mean that choices have to be made, the maintenance and care of existing features of high conservation value should always take precedence over the creation of new ones. However, the creation of new boundary features can add much to the wildlife and landscape value of a holding, and there are a number of measures that can be taken to speed up colonisation and encourage the development of wildlife diversity.

Among the most important of these is the appropriate choice of plant species and sites. Native species of grasses, wild flowers, shrubs and trees, especially those that are locally common, are most likely to establish successfully and to attract a variety of animal wildlife. Sites vary in their conservation potential. Although new features should not displace areas of high conservation value, they are more likely to develop diversity if they are sited alongside, or extend, existing enclaves of wildlife interest, such as ditches, woods and banks. Particularly promising sites, where soil fertility is low and there are plentiful sources of seed nearby, may be left to develop an interesting flora by natural regeneration. More fertile sites, and those with a history of troublesome weeds, need greater management intervention, for example the sowing of appropriate seed mixes. As with existing areas of habitat, new features should be protected from fertiliser and pesticide drift, and from other pollutants and excessive disturbance.

The cost of creating new features can be reduced by having a long-term plan that allows you to take advantage of any opportunities as they arise. For example, it may be a simple matter to create an extended grass margin alongside a hedge at the time that a grass ley is being ploughed up. Some options may easily be pursued by utilising existing resources of labour and machinery at slack times of the year. In the case of new roads, features such as mammal runs can be incorporated at minimal extra cost, provided that they are built into the design at an early stage.

One of the outcomes of an increasing public interest in environmental matters and government and European Community initiatives to limit food surpluses, is the provision of a range of grants to help with the creation of new boundary features (see Section 7.4 on grants in Chapter 7). Research into ways of encouraging wildlife in field boundaries has also been stimulated, resulting in the development of a number of novel options for farmland. Some have been designed specifically to reverse the decline in particular species, such as the grey partridge and certain arable weeds.

This chapter describes the main options currently being suggested for creating new boundary features. Some have been tried out on only a limited number of sites, so if you decide to adopt one of these it may be wise to introduce it on a trial basis at first, with careful monitoring and modifications to the management if necessary to achieve the desired result. You will find up-to-date information on the development of novel options in the farming press. Selection among the options available should be guided by the extent to which they match your own objectives and constraints, as described in Chapter 6 of the foundation book. The options finally chosen should be incorporated into your management plan, as illustrated at the end of this chapter for the case study farm.

6.1 Planting new hedges

Choice of site

The siting of a new hedge may be dictated by a particular need, for example to screen residential areas from the noise and visual intrusion of a new road, to mark a new boundary or to shelter soft fruit. It may be governed to some extent by constraints, for example the need to avoid field drains. If the choice of site is less restricted, then siting a hedge alongside existing boundary features, such as a ditch, bank or woodland edge, will enhance its wildlife value, and siting it alongside a public footpath or road will enhance its amenity value. A new hedge will be of greater value in landscape terms if it is publicly visible rather than hidden away in an isolated spot (and is more likely to attract grant aid from sources such as the Countryside Commission and local authorities). Other situations in which a new hedge might be of particular conservation benefit would include a hedge planted to divide an otherwise featureless expanse of arable land, to link isolated wildlife habitats, to continue an existing hedgeline or to recreate a former boundary.

Choice of species

Although hawthorn, blackthorn, holly, hornbeam and hazel are the most commonly used hedging plants, a wide variety of species can be used, as shown in Table 6.1. The choice is likely to be determined mainly by whether the hedge needs to be stockproof or not, the soil type and climate, and the intended form of management. In some situations, growth rate and tolerance of pollution, salt, wind or shade may be important. Information on the ideal site conditions and appropriate zones for all native trees and shrubs, including hedging species, is given in Table 6.1 and Figure 6.4 in *Woodlands* (see also Table 6.2 in the foundation book). A survey of the type of hedge that grows well locally will provide a further useful guide. Some species are poisonous to livestock (box, broom, cherry laurel, *Cupressus, Laburnum, Rhododendron* and yew) and so should be avoided on farmland.

The planting of a mixture of species makes for a more attractive hedge and increases its wildlife diversity. A predominantly hawthorn or blackthorn hedge can include up to 25% of other native species, such as dogwood, field maple, crab apple, hazel and some types of willow, without affecting its function as a stockproof barrier, although 10–20% is a more usual percentage, because of the additional cost involved.

Planting

Deciduous species are best planted after they have shed their leaves and before growth recommences in spring, that is to say normally between November and March and ideally before the end of December. Very wet or frosty periods should be avoided. Evergreens, such as holly, are better planted in milder conditions in May. Plants should be obtained from a reputable source, and should be well rooted and uniform in size (45–60 centimetres high). Plants supplied by nurseries are usually described as 1+1,

Table 6.1 Characteristics of native hedgerow species

Thorny or prickly species	
Blackthorn or sloe	Grows more slowly than hawthorn; withstands wind and salt spray; suckers freely so good for filling gaps; cuts and lays well.
Bramble and briars	Usually cut out of a hedge, but can help form a barrier in regions where hedge growth is stunted; dense bramble occurs on top of many turf hedges in exposed parts of the South-West.
Bullace	Closely related to blackthorn; bears thorn-like false spines which are actually short shoots; can be cut and laid.
Crab apple	Good for stock hedging but best used with other more impenetrable species. Should not be planted near orchards.
Gorse, furze or whin	Not a good hedge plant but withstands wind and salt spray, poor sandy soil and very dry sites; easily frosted so trim in spring or summer; cut to near ground level once plants become thin at the base; should not be laid.
Hawthorn, may, quick, whitethorn or thorn	Forms a good thorny hedge; grows in most soils and situations except at high elevation and in the shade; easily laid and tolerates regular hard trimming. Should not be planted near orchards.
Holly	Evergreen; slow growing but gives a sturdy, stockproof hedge; tolerates shade; good for filling gaps in thorn hedges; frost sensitive so best clipped or trimmed August–September; neglected hedges can be rejuvenated by cutting back hard or laying.
Cherry-plum or myrobalan	Fast growing but not as spiny as blackthorn and tends to become thin at the base; if maintained by trimming it should be clipped lightly late July–early August and again in December; may not respond well to hard cutting back and laying.
Smooth wood species	
Alder	Makes a good riverside hedge but does not tolerate acid peat or stagnant conditions; withstands hard cutting and laying; rarely eaten by stock or rabbits.
Alder buckthorn	Similar to alder in appearance and habitat requirements; no spines, open habit, so does not make a good hedge plant.
Ash	Quick growing; needs frequent laying if it is to be used as part of a stockproof hedge; difficult to cut when large.
Beech	Will grow on most well-drained sites; tolerates high altitudes and exposed situations; fairly shade tolerant; not a good stock barrier; retains its withered leaves if clipped in late summer, so useful as shelter; should not be cut back hard but can be laid.
Dogwood	Common on chalk and grows well on old red sandstones; forms a dense, tough hedge; suckers readily and seeds into pasture, so in places considered a weed; attractive red spring twigs and autumn foliage.
Elm	Badly affected by Dutch elm disease, so planting is not recommended at present; prefers heavy soils, withstands salt winds; stands hard cutting; suckers freely, so useful for filling gaps; can be laid; extensive root system can compete with crop and other hedge shrubs.

Field maple	Withstands salt air; usually found mixed with other hedging shrubs since it is not stockproof alone; can be cut and laid but is hard and brittle.
Hazel	Makes a strong, sturdy hedge plant on chalk and dry soils; traditionally managed by coppicing to produce pea sticks, hurdles etc.; little side branching so needs frequent laying if it must be stockproof.
Hornbeam	Stands hard cutting; grows well under dense shade; retains its dead leaves throughout the winter if trimmed in late summer; excellent firewood; can be used to add interest to a thorn hedge.
Oak	Usually occurs as a hedgerow tree, but can be managed as a hedge by cutting and laying.
Purging buckthorn	Found in hedges on chalk soil; not stockproof.
Scots pine	Used for narrow shelter-belts or screens; withstands dry conditions; can make a rough, quickly grown hedge if annual shoots are cut back by one-third to encourage side branches; tendency to thin at the base.
Wayfaring tree	Common in hedges on chalk soils; spreads into abandoned pasture; attractive, but too weak to make a good hedge plant.
Whitebeam	Quick-growing tree on chalk; adds interest to a hedge, but is not a good hedging plant itself.
Willow	Can make a rough hedge on wet ground, but not easily made stockproof; grazed by cattle; can be cut and laid but usually cut out of hedges before laying; useful for filling gaps on very wet ground; used as a windbreak in peaty areas.

2+1 and so on, the first number indicating the number of years that they have spent in the seedbed and the second the number of years spent in transplant lines; plants for hedging should be at least 1+1 and preferably 1+2. Plants will usually be supplied bare rooted, so care should be taken to keep their roots moist during transport, storage and planting. The site should be prepared by cultivating a strip about 60 centimetres wide along the proposed line of the hedge, incorporating some well-rotted farmyard manure if the soil is poor. If an earth ridge can be constructed on which to site the hedge, this will bring additional wildlife benefits. The plants should be set in the ground to the same depth as at the nursery. The soil around the roots should be well firmed in and any damaged or dead shoots should be removed. Setting the plants at an angle of 30° can result in a thicker hedge by encouraging additional shoots to develop.

The optimum spacing for hedgerow plants depends on the size and type of plants being used, the purpose of the hedge and the number of plants that you can afford. Planting can be either in a single line or in a staggered double row. If a hedge is to be managed by laying, plants can be spaced as far as 1 metre apart in a single row. At the other extreme, very dense planting, with the shrubs only 10 centimetres apart, can be used to create a stockproof barrier quickly without the need for laying. As a more general guide, for a stockproof hedge, to be managed by trimming, a staggered double row with 25 centimetres between plants within the row and 30 centimetres between rows is recommended. For areas without stock, or where a hedge is to be managed by laying, the distance between plants within the row can be increased to 50 centimetres and that between rows should be reduced to 15 centimetres.

Where stock are kept, appropriate fencing is needed to protect a new hedge from damage. Post and wire fencing usually provides adequate protection from browsing cattle, but more expensive woven wire fencing will be needed where there are sheep. Fencing should be sited far enough away from a hedge to ensure that animals cannot browse the tops of the young plants. If rabbits are a serious problem, chicken netting or individual tree guards may be justified for short stretches of new hedge. For more extensive areas, control of rabbits before a hedge is planted will probably be a more practical and far less costly option. Spring-planted hedges are less likely to suffer rabbit damage than those planted in the autumn. Rabbit damage may be reduced by planting a hedge into grass cover on top of a bank, although competition from the grass is likely to slow down hedge establishment, and may result in some plant losses.

More unconventional, low-cost options for establishing a hedge include growing your own plants from seed collected locally or simply fencing off a strip of ground and allowing natural scrub development to occur. Both options require considerable patience and plentiful sources of seed.

Early management

New hedges need to be protected from drought and excessive weed growth in the early stages if they are to establish successfully. In practice, careful use of approved herbicides in the immediate vicinity of the shrubs may be the simplest option for weed control, but mulches, where feasible, are likely to be less damaging to wildlife and will help conserve soil moisture. Natural mulches, such as hay, straw or wood chippings, applied after planting are ideal, but black polythene sheeting is an effective alternative. In some situations, polythene sheeting can be laid mechanically. Either plants can be inserted through slots in the sheeting or they can be planted first, immediately pruned to short, sharp stumps and the sheeting then applied and pressed into position so that the stumps emerge through it (see Box 6.1).

A further, but more risky, option for weed control might be to sow a low-maintenance grass mixture (for example of fescues and bents) at the time that a hedge is planted, to limit competition from more aggressive weeds and help establish a perennial grass sward.

If a hedge is to be managed by trimming, early pruning at planting or after the first year will encourage bushy growth. If it is to be managed by laying, the leading shoots should be allowed to grow unchecked to a minimum height of 2.5 metres, but side-shoots can be trimmed if necessary to control hedge width. (The method of establishing a hedge by applying polythene sheeting to pruned stumps is therefore inappropriate if a hedge will eventually be laid.)

6.2 Tree planting

Trees can add wildlife diversity and landscape interest to boundaries but, as with other new features, careful thought needs to be given to their siting. The shade that they cast may reduce the interest of existing features, such as flower-rich verges, and may cause gaps to develop in hedges. Trees grow more vigorously when supplied with fertiliser, so that tree planting may be inappropriate in verges where the main aim is to encourage fine grasses and a variety of wild flowers, achieved by keeping soil fertility low. Possible solutions to these potential problems may be to space trees widely or to

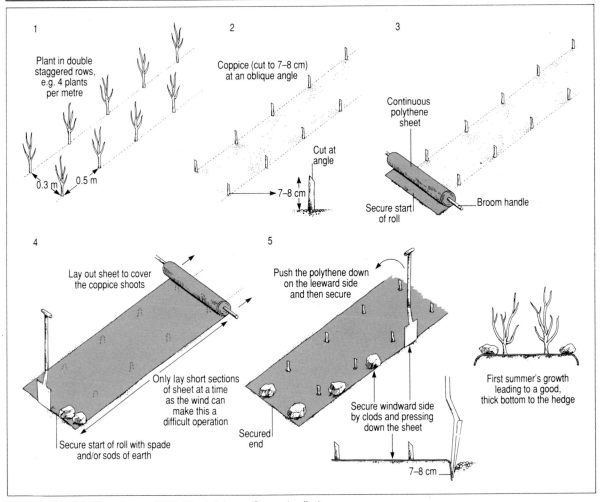

Box 6.1 Method of establishing a hedge using polythene mulch

1

Plant in double staggered rows, e.g. 4 plants per metre

0.3 m   0.5 m

2

Coppice (cut to 7–8 cm) at an oblique angle

Cut at angle

7–8 cm

3

Continuous polythene sheet

Secure start of roll

Broom handle

4

Lay out sheet to cover the coppice shoots

Only lay short sections of sheet at a time as the wind can make this a difficult operation

Secure start of roll with spade and/or sods of earth

5

Push the polythene down on the leeward side and then secure

Secured end

Secure windward side by clods and pressing down the sheet

7–8 cm

First summer's growth leading to a good, thick bottom to the hedge

(Source: Norfolk Farming and Wildlife Advisory Group leaflet)

plant them in groups with wide spaces between, to avoid the larger species or to restrict growth by pollarding or coppicing, or to restrict tree planting to more fertile areas of limited conservation value.

Tree species vary greatly in their conservation value. Native species, such as oak and hazel, support a far greater variety of wildlife than species that are relatively recent introductions to Britain, such as horse chestnut and European larch (see Figure 3.3 in the foundation book and Table 3.1 in *Woodlands*). For conservation purposes, native trees are therefore to be preferred. The choice of species should be related to the local climate, topography and soil type (see Table 6.1 and Figure 6.4 in *Woodlands*).

Methods of establishing hedgerow trees in existing hedges were described in Chapter 5. It is relatively easy to establish new trees at the time that a hedge is planted, because the competition from the hedgerow shrubs and ground vegetation will be less. Transplants in tree shelters generally survive better

Hedgerow trees

and grow faster than larger-sized plants. Native species, such as oak, ash, field maple, holly and beech, are recommended. Hedgerow trees should be planted at least 10 metres apart, but the spacing can be varied to avoid a formal and regimented appearance. Where hedgerow trees are unacceptable because of the shade that they would cast on arable crops, trees might be planted instead in groups in field corners, as shown in Figure 6.1.

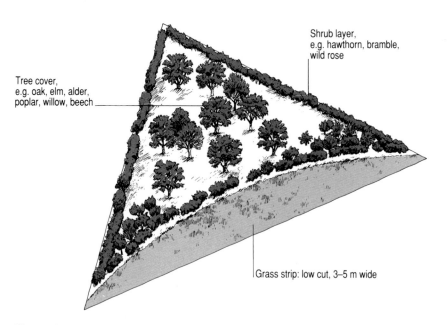

Shrub layer,
e.g. hawthorn, bramble,
wild rose

Tree cover,
e.g. oak, elm, alder,
poplar, willow, beech

Grass strip: low cut, 3–5 m wide

Figure 6.1   Tree planting scheme for a field corner

Shelter-belts

On farmland, single lines or narrow strips of trees can be used to form attractive, new boundary features, as well as providing shelter and some income from timber products. Mixed plantings of native coniferous and broad-leaved trees and shrubs are valuable in both wildlife and amenity terms and may attract grant aid (see Section 7.4 on grants in Chapter 7).

Some of the factors that affect the degree of shelter provided were mentioned in Box 4.1. To reduce wind speed most effectively, shelter-belts should not be more than 15–30 metres wide. They should be evenly permeable from top to bottom and so should include some shrubs and small, shade-tolerant trees, as indicated in Figure 6.2. If shelter is the main priority, a shelter-belt should be vertical at the edge; a graded, wedge-shaped edge will force the wind up and over rather than allowing it to filter through the trees. Stock and any damaging wildlife, such as deer and rabbits, need to be excluded by fencing to preserve the lower layers of vegetation and encourage natural regeneration.

Some fast-growing exotic species may need to be included to provide early shelter until slower-growing but longer-lived species take over. As a shelter-belt develops, some of the trees should be thinned to encourage a well-branched canopy. More detailed advice on the establishment and maintenance of trees is provided in Chapter 7 of *Woodlands*.

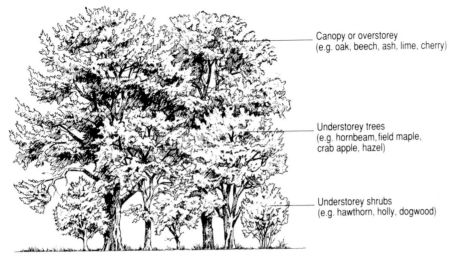

Canopy or overstorey
(e.g. oak, beech, ash, lime, cherry)

Understorey trees
(e.g. hornbeam, field maple,
crab apple, hazel)

Understorey shrubs
(e.g. hawthorn, holly, dogwood)

*Figure 6.2   Cross-section of a shelter-belt that includes some shrubs and small,
shade-tolerant trees to provide even permeability to wind*

6.3   Expanding field margins

A number of novel options have been suggested for expanding the width of
field margins to provide increased opportunities for wildlife, especially on
arable farms. These include conservation headlands, perennial grass
margins, arable weed margins and linear, grass-sown banks.

Conservation headlands

The recommendations for conservation headlands have been developed by
The Game Conservancy, initially in response to a decline in the numbers of
grey partridge on farmland. Because the recommendations centre on
reduced pesticide use at the crop edge, conservation headlands benefit not
only gamebirds but a variety of wildlife, including some of the rarer arable
wild flowers, such as shepherd's-needle, weasel's snout and night-flowering
catchfly, useful pest predators, such as earwig, ground beetles and rove
beetles, small mammals, especially woodmice attracted to the weed growth,
and butterflies and pollinating insects (attracted by the nectar).

The Game Conservancy's research showed that, in the first few weeks of life,
grey partridge chicks mainly feed on insects, such as sawfly larvae, ants,
plant bugs and leaf beetles, and that these in turn are encouraged by the
presence of arable weeds, such as knotgrass, black bindweed, mayweed and
charlock. The widespread use of herbicides and insecticides has therefore
markedly affected the chicks' food supply (see Box 3.2 in the foundation book).

To encourage a moderate growth of arable weeds at the edge of cereal crops
where the chicks feed, herbicide use on the outer 6 metres or so of a crop (the
conservation headland) is restricted to the selective control of serious weeds.
Broad-leaved weed herbicides are avoided unless absolutely essential.

Insecticides are allowed for the control of Barley Yellow Dwarf Virus (BYDV) in the autumn, but none are used after 15 March, when the insects that the chicks feed on begin to emerge from their overwintering sites. Most fungicides can be used as normal, although one or two may have some insecticidal effects. A summary of The Game Conservancy's recommendations is shown in Figure 6.3, which includes advice on management beyond the headland area.

The exact width of a conservation headland is determined by the width of the machinery used in a crop. Since the chicks rarely venture very far into a crop, a headland 6 or 8 metres wide is sufficient for game purposes. If the incursion of serious weeds is a problem, sterile strips of ground 0.5–1 metre wide can be used as a barrier between the edge of the field and the crop, although this has no benefit for wildlife apart from providing a dusting area for birds. These strips are kept clear of vegetation by cultivation two or three times a year or by careful application of herbicide in early March.

Yield reductions in conservation headlands are normally relatively small (5–10% yield loss compared with conventionally sprayed headlands), except where there is poor crop establishment, when a wet summer encourages heavy weed growth or when fungicides used on the rest of the field are applied in a tank mix with other pesticides, and therefore cannot be applied to the headland area. Some additional costs may be incurred if extra drying and seed cleaning are necessary.

The choice of sites for conservation headlands is important to their success. Drier sites are preferable to wet areas, both because partridge prefer dry ground and because serious infestations of cleavers and barren brome are less likely to occur. Conservation headlands in winter-sown crops are less likely to suffer from weed competition than those in spring-sown crops. The potential loss in income as a result of reduced grain quality in headlands can be minimised by avoiding crops for which quality is particularly important, such as malting barley, or by harvesting the grain from the headland area separately and using it for animal feed.

In some areas, conservation headlands on arable land qualify for grant payments (see Section 7.4 on grants in Chapter 7).

Perennial grass margins

The establishment of perennial grass margins around the edges of a field can be used to control the growth of problem annual weeds, by creating a dense sward that makes it difficult for the weeds to re-establish from seed. Grass margins give additional protection from operations in the field to other boundary features, as well as providing easy access for machinery and crop inspection, convenient areas for siting beehives to improve pollination in crops, such as linseed and oilseed rape, and attractive footpaths and bridleways.

Benefits to wildlife   Regular mowing of grass margins that have a diverse flora favours fine grasses, such as red fescue, sheep's fescue, crested dogstail and sweet vernal grass, and wild flowers, such as self-heal, knapweeds, field scabious, ox-eye daisy, meadow vetchling and birds-foot-trefoil. This flowery grass sward attracts butterflies and bees. Less intensive management, in which the margins are only cut to control invasion by scrub, allows coarse, tufted grasses, such as cocksfoot, false oat-grass and tufted hair-grass, to flourish. The resulting thatch of old grass stems and roots, especially if relatively free

THE GAME CONSERVANCY'S FIELD MARGIN

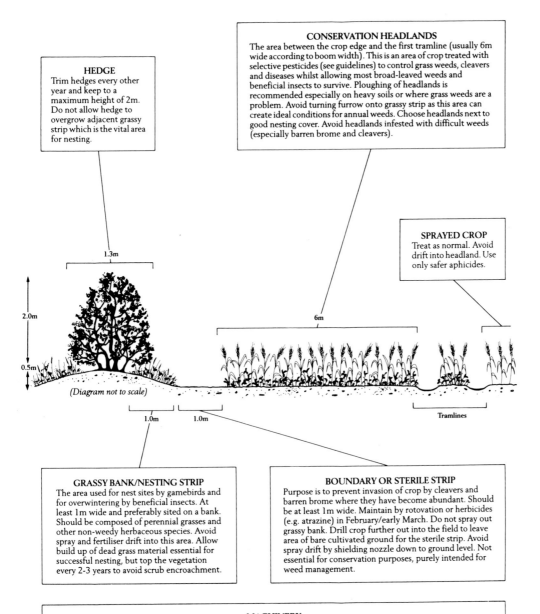

HEDGE
Trim hedges every other year and keep to a maximum height of 2m. Do not allow hedge to overgrow adjacent grassy strip which is the vital area for nesting.

CONSERVATION HEADLANDS
The area between the crop edge and the first tramline (usually 6m wide according to boom width). This is an area of crop treated with selective pesticides (see guidelines) to control grass weeds, cleavers and diseases whilst allowing most broad-leaved weeds and beneficial insects to survive. Ploughing of headlands is recommended especially on heavy soils or where grass weeds are a problem. Avoid turning furrow onto grassy strip as this area can create ideal conditions for annual weeds. Choose headlands next to good nesting cover. Avoid headlands infested with difficult weeds (especially barren brome and cleavers).

SPRAYED CROP
Treat as normal. Avoid drift into headland. Use only safer aphicides.

GRASSY BANK/NESTING STRIP
The area used for nest sites by gamebirds and for overwintering by beneficial insects. At least 1m wide and preferably sited on a bank. Should be composed of perennial grasses and other non-weedy herbaceous species. Avoid spray and fertiliser drift into this area. Allow build up of dead grass material essential for successful nesting, but top the vegetation every 2-3 years to avoid scrub encroachment.

BOUNDARY OR STERILE STRIP
Purpose is to prevent invasion of crop by cleavers and barren brome where they have become abundant. Should be at least 1m wide. Maintain by rotovation or herbicides (e.g. atrazine) in February/early March. Do not spray out grassy bank. Drill crop further out into the field to leave area of bare cultivated ground for the sterile strip. Avoid spray drift by shielding nozzle down to ground level. Not essential for conservation purposes, purely intended for weed management.

MACHINERY
A specially designed sprayer is now available which can selectively spray a six-metre strip along the headland while treating the main crop with standard chemicals. Each part of the machinery is independent of the other, thus saving the need for a separate run along the Conservation Headland.
While spraying sterile strips it is vital to prevent drift into crop and hedge bottom. A very useful device, which applies the chemical safely and accurately from the tractor, has been designed for this purpose.
For further information on these two pieces of equipment contact the Conservation Headlands Field Officer.

Figure 6.3   The Game Conservancy's Field Margin. (Source: The Game Conservancy leaflet)

from disturbance, provides excellent cover for invertebrates, such as beetles and snails, as well as mammals, such as field vole, shrews and hare, and ground-nesting birds, such as pheasant. It also provides good hunting ground for kestrel and barn owl. Wide, damp grassy stretches situated away from roads and mown once every two years are especially good territory for barn owls.

Establishment

Options for establishing perennial grass margins include allowing natural regeneration on bare ground, sowing a conventional grass ley, sowing a mixture of fine grasses and wild flower seeds, fencing off an area of pasture or leaving a grass margin when ploughing up existing leys. Natural regeneration relies almost entirely on the presence of seed sources close by, either in the field or the field margin, so that for this option to be successful, the local flora must be relatively rich. If this is the case, a great variety of species may colonise in the first year, although the more competitive problem weeds, such as wild oat and black-grass, may predominate. These undesirable weeds can be discouraged by mowing them as they flower to prevent seed formation, and by spot treatment with herbicide if necessary, although most annuals are likely to decline naturally in the second year. Once any problem weeds are under control, the frequency of mowing can be reduced.

Where there is a serious weed problem (more likely on heavy and fertile soils), a sown grass margin is preferable. A conventional grass seed mixture, containing cocksfoot, false oat-grass and ryegrass, can be used, or a mixture of slow-growing grasses can be applied if minimum maintenance is desirable. Seed mixtures containing native fine grasses and locally occurring wild flowers, although attractive and beneficial for wildlife, are considerably more expensive, and only likely to succeed on infertile soils and with regular mowing. The cost can be limited by selecting mixtures containing only the more showy and readily established wild flowers, such as ox-eye daisy and knapweeds. A ratio of 70–80% grass seed to 20–30% broad-leaved plants is recommended. If there is old pastureland nearby, this can be used to provide seed, or cut green hay or the chaff from baled hay taken from the pasture can be spread in the grass margin as a seed source.

In established grass, diversity can be encouraged by scarifying the soil surface with a spiked harrow or metal rake, light poaching by grazing animals, slot seeding with a wild flower mixture or planting out pot-grown wild flowers or flower-rich turves. As already mentioned, a greater diversity of plants is likely to occur on sites where soil fertility is low or readily depleted. Regular mowing and removal of cuttings prevents smaller plants being smothered and may slowly reduce soil fertility. If a coarse grass margin is to be established on pasture, a fence will be needed to prevent the area being grazed.

Grass margins may be as little as 0.5 metres wide, although the wider they are the greater the benefit for wildlife, provided that they are taken from the cropped area and so extend the existing boundary habitat. In practice, the width is normally determined by the width of machinery that will be used along them. In some circumstances, the establishment and maintenance of wide grass margins can qualify for grant aid (see Section 7.4 on grants in Chapter 7).

Surprising at it may seem, arable weeds are now one of the most threatened groups of plants in the British flora. Chemical weed control, especially the use of broad-spectrum and persistent herbicides, more efficient methods of seed cleaning, the change from spring to autumn sowing (with a consequent reduction in spring-cultivated seedbeds) and increased fertiliser use have all contributed to their decline. The establishment of arable weed margins, also known as annual flower strips, uncropped wildlife strips and wildlife fallow margins, is designed to encourage and preserve some of the more attractive arable weeds.

Arable weed margins are also likely to benefit other groups of wildlife: plant-feeding insects, such as those that live on fumitory, *Polygonum* and common poppy; butterflies attracted by the nectar of plants like charlock and field pansy; insectivores, such as spiders and shrews; ground-nesting birds, such as skylark, lapwing, stone curlew and pheasant; seed-eating birds, such as thrushes, finches, buntings and pipits; herbivorous mammals, such as the brown hare.

Arable weed margins are cultivated either every year or every other year, depending on the species to be encouraged. It is normally sufficient to disturb the top 6–15 centimetres of soil with a cultivator, rotavator or power harrow. A flail mower can be used first to reduce the bulk of rank growth if necessary. Deep ploughing should only be needed if the ground is infested with barren brome or black-grass, and should be followed by cultivation to produce a reasonable seedbed. Alternatively, a herbicide can be used to kill aggressive weeds at least two weeks before cultivation. Spot treatment with herbicide can also be used to control undesirable weeds as the margin flora develops. As with other conservation areas, fertiliser and herbicide drift from the cropped area should be prevented. Mowing of arable weed margins is undesirable, as it encourages tillering (the formation of side-shoots) in annual grasses and favours perennials at the expense of annual flowers. Machinery should be kept off margins from the beginning of April to the end of July, to avoid disturbing ground-nesting birds.

The flora that develops will be influenced by the time of year when the ground is disturbed. Spring cultivation encourages the germination of weeds such as fool's parsley, black nightshade and fat-hen, whereas autumn cultivation encourages speedwells and mayweeds. On dry soils, species such as parsley-piert, henbit dead-nettle and field pansy occur, while on wetter soils, redshank and orache are common. Where the aim is to encourage a particular local rarity, management should be matched to its germination requirements if these are known. For example, shepherd's-needle germinates in the autumn and has poor seed dormancy, so that regular autumn cultivation is essential for its survival.

As with other new boundary features, arable weed margins are best sited adjacent to areas with a diverse flora and away from sources of problem weeds. Light, free-draining soils generally give the more successful results. Relatively few types of arable weed are likely to germinate on heavy soils or where an area has previously been down to pasture for any length of time. The margins are especially valuable in localities where rare arable weeds are known to occur but are in danger of being lost. The width of a margin is not important, so is most conveniently based on multiples of the width of cultivating machinery. Weed populations will often be highest close to the edge of a field. Wide, uncropped wildlife strips taken from arable fields may qualify for grant aid (see Section 7.4 on grants in Chapter 7).

Linear, grass-sown banks

Recent research has examined ways of encouraging large numbers of pest predators within cereal crops, by creating linear, grass-sown banks across the middle of fields. The banks are easily constructed, and if sown with grasses and protected from spray drift can encourage the rapid build-up of very large numbers of predators within a short time. The banks enhance the penetration of predators into the crop in the spring and so help with pest control. They also provide new boundary features of wildlife interest within arable fields at minimal cost and in a short space of time. For agricultural purposes, linear banks have the advantage of being easily removed, although for most conservation purposes, the more permanent a feature, the greater its value to wildlife.

6.4 Building stone walls

Conservation guidelines for the building of new stone walls are similar to those already described in Chapter 5. For example, they should be built in the local style using local sources of stone wherever possible. Dry stone walls are more valuable in conservation terms than those sealed with mortar, since they provide more crevices for wildlife to exploit. The use of mortar in the coping layer may weaken rather than strengthen a wall, by reducing its flexibility as it settles and moves. The basic procedure for building a typical dry stone wall is illustrated in Box 6.2, but it is a skilled job for which practical training or professional help will probably be needed.

Box 6.2 Building a dry stone wall

The procedure described here relates particularly to free-standing dry stone walls characteristic of areas such as the Pennines (see Box 2.1), but many of the same general principles apply to other types of dry stone wall.

Walls should be about 1.2 metres high for cattle and about 1.6 metres high for sheep. A **batter frame**, or template, can be used to ensure that a wall is built to the correct profile and that the courses are level. The batter, or taper, of a wall depends on the type of stone being used.

Uneven-sized stones need to be built with a greater batter to provide a stable wall. The best-shaped stones are used as **face stones**, larger stones being used near the bottom and smaller stones near the top. The smallest, most angular stones are used as **fillings**. **Through stones**, which run across the full width of the wall, are incorporated into the wall at regular intervals to give it greater strength. Many walls are topped with a layer of **cope stones**, which help bond the two faces of a wall together and weigh down the courses, helping the wall to settle.

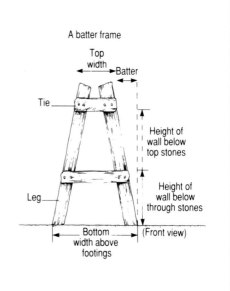

To build the wall:

1   The foundation trench is marked out, cleared and dug.

2   The stones are roughly sorted according to their intended use (as face stones, through stones or coping stones) in piles either side of the trench, leaving a space in which to work.

3   The foundation stones are laid in two parallel rows in the trench and the space in the middle is packed with fillings.

4   The batter frame is propped in place at the point where it is judged that the day's work will finish, and guidelines (e.g. of nylon twine) are stretched at a height of about 0.3 metres between the frame and the existing section of wall (two frames, one either end, are needed initially).

6   The guidelines are raised to the first line of through stones and the courses continued up to this height, taking extra care at this stage to ensure that the top is level.

7   The through stones are built into the next course, at intervals of less than 1 metre provided that sufficient suitable stone is available.

8   The guidelines are raised again, either to the level of the next layer of through stones if there is to be one or to the coping level, and building is continued as before. Placement of through stones should be staggered if there is more than one line, to give extra strength to the wall.

Step 4 Positioning the batter frame and guideline

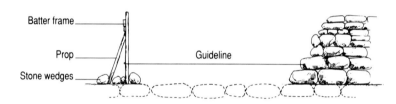

Batter frame — Prop — Stone wedges — Guideline

5   The courses are built up to this level, following several basic rules:

(i)   the biggest face stones should be placed at the bottom;

(ii) joints should be staggered, to provide a strong bond between successive courses;

(iii) the longest side of each stone should run into, rather than along, the wall for greater strength;

(iv) face stones should be positioned with their tops level, or tilted slightly downwards towards the outside, to prevent rainwater being channelled into the middle;

(v)   the middle should be packed solidly with fillings at the same time as each course is built;

(vi) each course should be set in slightly from the previous one to achieve the required batter.

9   Cope stones are not usually put into position until the whole length of the wall is complete. Unfinished sections are left with stepped courses so that the join between consecutive stretches is firm. In some areas, flat cover stones are used for the final course, covering the width of the wall if possible. These help to shed rainwater and provide a firm base for the cope stones.

6.5   New transport routes

Major new road schemes present both threats and opportunities for conservation. They may damage areas of high conservation value, but they may also provide large tracts of land where wildlife can exist relatively undisturbed. Their construction should be preceded by an environmental appraisal, which first considers whether they are justified and then determines their location, design, management and ways of minimising the impact on the environment at all stages.

Wherever possible, sites of high conservation value should be avoided and the least damaging route selected. When a route divides a valuable wildlife site, mobile species will be killed by vehicles and, since their territory is split, their breeding success will be reduced. Some species, such as wading birds, may move away as a result of disturbance, while others may be lost as a result of changes in the drainage pattern.

Where routing a road through valuable conservation areas is unavoidable, it may be possible to take special measures, such as rescuing turf or moving coppiced stools like sallow, to help create new habitats elsewhere. However, such procedures are normally costly and of limited value, so that existing features of wildlife interest, such as hedges, ditches, ponds and trees, should be retained and incorporated into the design of a road wherever possible. Care should be taken during road construction to avoid dumping materials and soil, and parking equipment and machinery, on areas of conservation value that are to be retained. Where a wildlife survey has revealed obvious signs of animal activity, such as the presence of badger setts, near the proposed line of a road, several features can be incorporated into the design of the road to improve the chances of continuity. Where a **culvert** is installed to channel water underneath a road, it can incorporate a ledge at the side above the water line to allow animals, such as badgers, foxes and deer, to reach the other side of the road in safety. Alternatively, additional culverts can be installed above the ground water line specifically for animals to use (Figure 6.4). Where a road bisects an established run, chain link or sheep fencing may be used to fence off the road at the normal crossing-point and guide animals to the entrance of a new culvert. New bridges that carry a road across relatively undisturbed areas, such as rivers and flood meadows, can incorporate crevices or hollow concrete blocks underneath to provide roosting areas for bats.

Many of the options and guidelines already discussed in this chapter for creating new boundary features also apply to the establishment of vegetation on new roadside verges. Trees will grow more vigorously if planted on the more fertile areas of verge with a reasonable depth of topsoil. The species chosen should normally be typical of those that grow well in the area, although where site conditions are poor or where a rapid screen is needed, pioneer or nurse species, such as alder, willow, birch or larch, which may not be local, are sometimes used. The range of native trees and shrubs that can be used to add interest to tree-planting schemes, and the areas where they are likely to grow well, are shown in Table 6.1 and Figure 6.4 in *Woodlands*.

Where a grass verge is planned, the depth of topsoil can be limited to 50 millimetres, to restrict fertility and encourage the development of a diverse sward. The options for establishing grassland are similar to those already

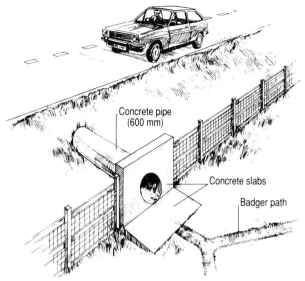

Badger tunnel under a road on an embankment with exclusion fencing

Culvert under a major road showing raised pathway for use by badgers

Figure 6.4   Badger underpasses. (Source: Harris et al., 1988)

given for perennial grass margins on farmland. In some areas with light, infertile soil, such as on chalk or greensand cuttings, and where there are adjacent areas of wild flowers, topsoil can be omitted altogether and natural colonisation allowed to occur. Where the local flora is poor, or a quick result is needed, a seed mixture of native grasses and locally common wild flowers can be sown. On steep-sided cuttings, the seed is sometimes sprayed onto the banks in a mixture of bulky organic material and water (**hydroseeding**) or applied in seed-impregnated matting, to help retain the seed in place and stabilise the soil. Ideally, the flower seed should be collected from a local source and a record kept of the species sown. Costs can be reduced by sowing small areas with a wild flower mixture to act as a seed source for colonising adjacent areas sown to grass. Trials are under way to develop conservation seed mixtures of short statured and slow-growing plants for

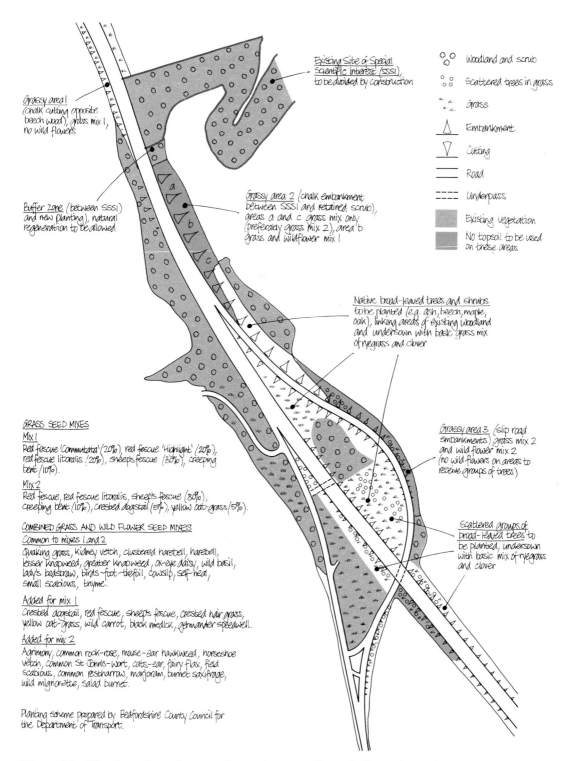

Figure 6.5   Planting scheme for a new bypass incorporating a chalk cutting and embankment (Barton-le-Clay bypass, Bedfordshire)

roadside verges on particular soil types. An example of a planting scheme for a new bypass, which incorporates a chalk cutting and embankment, is shown in Figure 6.5.

The new railway link between the Channel Tunnel and London provides a now-rare example of the need to establish new lineside boundary features. Similar guidelines and options to those for trunk road and motorway verges, embankments and cuttings are likely to apply. Tree planting may be best restricted to embankments where there is less risk of trees, branches and leaves falling onto the track. As mentioned in Chapter 3, it is possible to incorporate badger crossings on lines that are to be electrified in areas that show signs of heavy badger activity. Gaps of 3–4 metres can be left in the conductor rail at the site of known runs, provided that they are spaced at least 50 metres apart. As with roads, fencing can be used alongside a track to guide the animals to a crossing-point.

6.6 Deciding on options for new boundary features

Consider as wide a range of options as possible before narrowing down your choice. Start with any of the options already described which apply to your site, and if necessary search for further ideas by consulting trade magazines or friends and neighbours, seeking expert advice or looking at similar sites elsewhere. Choose among the options available by ranking them according to how well they match your objectives and constraints. Remember that the maintenance of existing boundary features should normally take precedence over the creation of new ones, if a choice has to be made. Do not limit your choice to those options that can be implemented immediately; some can be added to your long-term plan, to be implemented when the opportunity arises or when circumstances change.

6.7 Exploring and choosing options for the case study farm

This section covers the options for both existing and new boundary features on the case study farm. Bear in mind that the farm was selected as an example partly because of the diversity of boundary features that already exist and partly because of the landowner's longstanding interest in wildlife. As a result, many conservation measures are already being put into practice, and where these are having the desired effect the existing management will be continued. However, a number of options for further improvements were identified by the adviser as a result of the integrated assessment, and these were suggested for the landowner and farm manager to consider. Those options that they accepted were incorporated into a conservation management plan, as shown in Figure 6.6.

In reading this section, notice the important part played by personal perceptions in the selection of options. In particular, the farm manager is concerned about how the farm will look to other people, as already acknowledged in Figure 4.3. Because he recognises that this is a constraint based on individual opinion rather than economic necessity, some of the options concerned are kept open for further discussion.

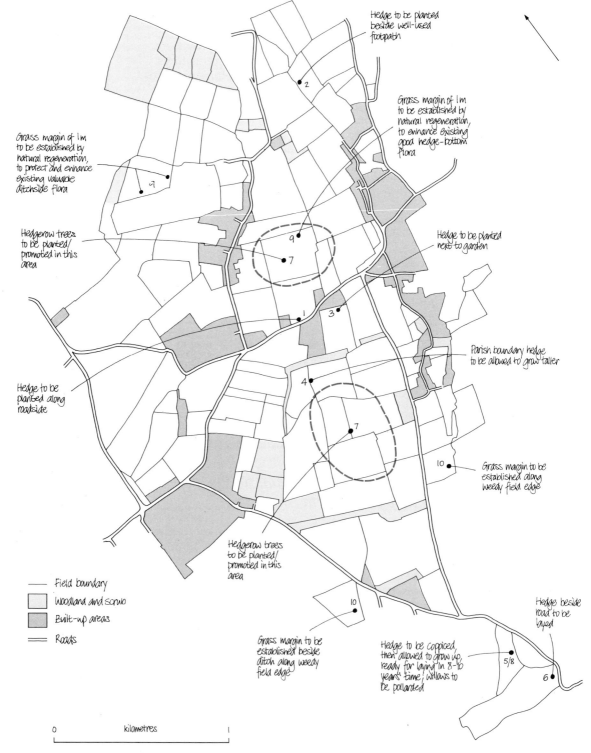

Hedge to be planted beside well-used footpath

Grass margin of 1m to be established by natural regeneration, to enhance existing good hedge-bottom flora

Grass margin of 1m to be established by natural regeneration, to protect and enhance existing valuable ditchside flora

Hedgerow trees to be planted/ promoted in this area

Hedge to be planted next to garden

Hedge to be planted along roadside

Parish boundary hedge to be allowed to grow taller

Grass margin to be established along weedy field edge

Hedgerow trees to be planted/ promoted in this area

Field boundary
Woodland and scrub
Built-up areas
Roads

Hedge beside road to be layed

Grass margin to be established beside ditch along weedy field edge

Hedge to be coppiced, then allowed to grow up, ready for laying in 8–10 years' time; willows to be pollarded

0   kilometres   1

Figure 6.6   Summary conservation management plan for boundaries on Kemerton Estate

Planting new hedges

There are several field boundaries on the farm that at present have no hedgerow and which would benefit from hedge planting. This applies particularly in the case of the large, grass field on the hillside, referred to in the landscape assessment. A new hedge planted here, beside the wire fence that already divides the field, would have a significant impact on the landscape as seen from Bredon village, and would have little agricultural cost once established. It would form a valuable corridor for wildlife, by connecting the young woodland and scrub above the field with the thick hedges below it. However, the farm manager commented that because of the steepness and undulations of this field it would not be possible to use a hedge trimmer on any hedging planted. In addition, the landowner felt that this large, open field on the hillside was an attractive feature. This option was therefore rejected.

The planting of new hedges along three other field boundaries was discussed and agreed. One hedge will be planted along a roadside, to give a more distinct and attractive boundary to a field where only a weedy grass margin now exists. A second hedge will be planted on a visually prominent boundary on the flanks of Bredon Hill, beside a well-used footpath. A third hedge will be planted between a garden and a field. In each case, the hedge will be planted on the field side of the boundary, leaving the existing grass margin intact. The hedges will be planted at a spacing of six plants per metre, with a mixture that mimics that in old, local hedges: 80% hawthorn, 7% each of dogwood and blackthorn, and 2% each of field maple, hazel and dog rose. The cost of planting the new hedges and of putting up the necessary protective fencing is currently eligible for a grant from the Ministry of Agriculture, Fisheries and Food (MAFF).

New management of existing hedges

At present, all the hedges that are trimmed are cut annually. Alongside roads this is necessary to ensure good visibility for traffic. For internal hedges, however, the adviser suggested that cutting might be less frequent to allow the shrubs to flower and produce fruit on the second year's growth. The landowner's reaction to this suggestion was positive, particularly as it would save money and time, but the farm manager felt that the extra year's growth would give the farm an untidy and unkempt look. He also felt that the farm's hedge trimmer, and the tractor that it is mounted on, might not be powerful enough to give a clean cut to the thick, woody stems. This option will be the subject of further discussion.

Around the arable fields in the southern half of the farm, many of the hedges are low and relatively thin. To increase their width and perhaps their height, it was proposed that the hedges might be allowed to grow gradually wider. Around the sand and gravel quarry, many of the hedges have been allowed to grow tall as a screen. The landowner and farm manager agreed that other hedges might be allowed to grow taller, for example the old parish boundary hedge, which has been identified as an important wildlife corridor. It has a track on its western side and runs north–south, so crop shading should not be a problem.

Where the low hedges have developed large, woody stumps, coppicing was proposed as an option. Any large gaps could be planted with hawthorn whips. It was agreed that one such hedge should be coppiced next winter. This work will be eligible for a MAFF grant.

Around some of the permanent pasture, the hedges have been allowed to grow up unchecked to give shelter to the stock. Some long-term management has been carried out on those hedges that have become gappy or top heavy, usually by reducing the height of the hedge with a shape saw. In addition, about 500 metres of hedge are laid every winter by farm staff, who were trained on an Agricultural Training Board course held on the farm. A MAFF grant helps to pay for the work, which is to be continued: a hedge will be laid next winter. However, many of the more easily laid hedges have already been tackled, so that it may be necessary to coppice some of the others before laying them in about eight years' time.

Encouraging hedgerow trees

The importance of hedgerow trees in the landscape, especially since the loss of the elms during the 1970s, has already been stressed in the landscape assessment. Ash, wild plum, oak and sycamore have been allowed to grow up from many of the low hedges on the farm, especially along roads, and have already made a significant impact. However, some of the trees that have developed from hedgerow stools are not very vigorous. The adviser therefore suggested that some new standard trees might be planted into hedges in two areas that at present have few hedgerow trees and relatively low hedges. Pedunculate oak, field maple and holly would all be suitable. There was some difficulty in identifying appropriate sites because of electricity lines, and because the management of one of the hedges is shared with a neighbour, but it was agreed to consider the planting of several new trees in the future.

Pollarded trees

Although there are a large number of old, pollarded willows on the farm, regular pollarding has lapsed in the last few years. As a consequence, many of the older pollards are starting to split and rot. During the winter of 1986–7, several of the willows were pollarded by the conservation warden, with the help of a grant from the county council. In 1987, others were pollarded by the water authority during regular maintenance work on Carrant Brook. A longer-term programme of pollarding was suggested at the time and is now gradually being implemented.

Field edge management

The width of the field margins at present varies from field to field, and is determined arbitrarily by the ploughman. In some fields, there is a grassy margin of 1.5 metres, while elsewhere fields are ploughed right up to the hedges. In view of the low agricultural productivity and high conservation value of the area close to a hedge, the adviser suggested that a margin of at least 1 metre should be left unploughed around all arable fields (1.5 metres where there are footpaths). The farm manager said that to attempt to switch to a 1-metre uncultivated margin around every arable field in the same year might be taking on too much. He envisaged many of the margins becoming an uncontrolled mass of cleavers or other unacceptable weeds. It was therefore agreed to target four field edges in the first year. Sites were chosen on the basis of their importance as wildlife corridors or because of their existing valuable flora of non-weed species.

The adviser suggested that, when a field is next cultivated, a deep furrow should be ploughed in towards the crop, to establish a clean break between the grassy edge and the crop. Disturbance of the field edge habitat, either physical or chemical, should be avoided. The farm manager agreed with this in principle, but said that in practice creating a perfectly clean break in this way was impossible. The use of a narrow, sterile strip between a boundary and a crop (created by spraying out the vegetation) has been designed to help overcome this problem. Ideally, this protective strip should not be bare soil, but consist of coarse, perennial grasses, such as cocksfoot or false oat-grass. In future, where grass leys are being ploughed for arable crops, the existing vegetation can be left around the field edges. In other areas, the sowing of grass along some field edges may be considered.

The drift of fertiliser into the hedge bottoms is being prevented by the use of a tilt mechanism on the fertiliser spreader, which is engaged when the headlands are being fertilised.

Conservation headlands

Conservation headlands have been used on the farm since 1987, as part of The Game Conservancy's Cereals and Gamebirds Research Project. The location of the headlands is chosen by the farm manager and the conservation warden together, to minimise agricultural problems and maximise the conservation benefits. This means that headlands infested with problem weeds, such as barren brome, black-grass, annual meadow-grass, couch grass and cleavers, are avoided. The landowner sees the main benefits as the conservation of the uncommon arable weeds found on the farm, rather than the increase in wild game populations. Headlands with a diverse flora of uncommon arable weeds are therefore targeted. There is a degree of conflict in that many of the interesting arable flowers occur on the light, limestone brash soils, which grow quality malting barley (weed seed contamination affects the quality of the sample). In practice, the contamination of weed seed is not usually significant. The main problems are for farm staff to remember not to spray certain pesticides and for the farm manager who has to be aware of any developing pest or weed problems. To overcome these, the location of conservation headlands is marked with large signs to make recognition easier. Species diversity and the abundance of the weed population in the conservation headlands is monitored by the conservation warden, to help guide site selection for the headlands in future.

Grass margins

Existing grass margins are mown once a year before harvest, with a pasture topper or flail mower. Over the last few years, several new tracks have been created. The Oathill Track previously ran along a steep bank beside a hedge, and has been moved out towards the field onto flatter ground. The bank now has a diverse, calcareous flora, including several locally uncommon species, such as common calamint, while the new trackway has gradually developed a grassy flora.

New grass tracks beside two other fields have been sown with a mixture of hardy prostrate perennial ryegrass and finer grasses, such as fescues, bents and meadow-grasses. The intention is that the ryegrass will dominate in the wheel tracks, while the finer grasses will fare better next to the hedges and will allow colonisation by wild flowers. The tracks have been mown several times a year, to encourage grass establishment. The conservation warden is monitoring the development of the flora.

Where footpaths run along the edges of fields, a width of 1.5 metres has been established during the last few years to meet the requirements of the 1990 Rights of Way Act. To encourage walkers to use the paths and to allow a grass sward to develop, the footpaths have been mown in spring and again before harvest, with a narrow flail mower mounted behind the farm's mini-tractor .

There is little agricultural need for further tracks or pathways around fields, and the cost (in terms of loss of cropped area) of new conservation grass margins wider than 2 metres means that it is unlikely that more of these will be created, unless as part of a compulsory set-aside scheme. However, where the existing flora of a hedge bottom or field edge is poor (containing problem agricultural weeds), there is scope for establishing narrow grass margins. The farm manager felt that this might be most appropriate along field edges with particularly enriched and weed-ridden soil, as the vigorous growth of the grasses would suppress annual weeds. Two sites were chosen as being suitable for the first year. Sites likely to encourage walkers were avoided, since there is already considerable public access over the farm.

To establish the grass margins in as weed-free an environment as possible, the existing weedy flora will be carefully sprayed out during the summer and seedbeds prepared with the rest of the fields in the autumn. The flush of germinating weeds can then be sprayed off again with a systemic herbicide. A grass mixture of cocksfoot, Yorkshire fog, brown top bent and perennial ryegrass will be sown onto 2-metre strips without disturbing the soil again (since this would encourage another flush of weeds). If possible, the strips will be mown with a flail mower mounted on the back of the farm's mini-tractor. If necessary, the width of the grass margins can be reduced to 1.5 metres after a year or two. The margins will add an important wildlife habitat to the field edges, and should reduce the invasion of agricultural weeds from the hedge bottoms.

The adviser suggested that the introduction of wild flowers into existing margins of little conservation value, or into the new grass strips, might also be considered. Young plants or seed could be used, but any existing vegetation would need to be removed first, unless a slot seeder could be hired to insert seeds into existing swards. Suitable wild flower species include red campion, self-heal, hedge woundwort, common knapweed and common St John's-wort on the clays and wild carrot, greater knapweed, hedge-bedstraw and musk mallow on the limestone. The landowner was unhappy about introducing wild flowers of unknown provenance into field boundaries on the limestone soils, since local flowers are likely to colonise naturally from nearby seed sources given time and correct management. On the heavy clay land, wild flowers are less common and natural colonisation takes longer, so he agreed to give further consideration to the sowing of wild flower seeds on a few of these sites.

Arable weed margins

The first arable weed margin on the farm occurred by accident. During the laying of a gas main, the gas board disturbed a 5-metre strip on light, gravelly soils at the edge of an arable field. The flora that developed on this uncropped strip was remarkably diverse, with up to 13 species of broad-leaved weeds per square metre and no problem weeds. The strip produced an attractive array of flowers (to non-farmers at least!) and was used by many insects and songbirds. Since then the margin has been cultivated annually with the rest of the field, but left uncropped. Unfortunately, it has gradually become dominated by perennials, such as cocksfoot and broad-leaved dock, which have seeded in from the hedge bottom. Ploughing of this strip, and perhaps a year under crops, may overcome this problem. Where arable weed margins have been tried elsewhere on the farm, they have been less successful. One strip on heavy clay soil was a notable disaster, with cleavers swamping all other plants.

The location of new arable weed margins is limited by the distribution of diverse, non-problem, annual weed communities. These tend to occur in relatively few field edges on the lighter soils. Although the siting of arable weed strips may vary from year to year, there is probably not great scope for increasing the number that already exists on the farm. The one exception is in the area of the gravel quarry. During excavation, an interesting range of arable weeds has grown on the disturbed soil, including four different species of poppy. It is therefore intended to include arable weed strips in the plan to turn this area into a wetland nature reserve once excavation is complete.

Ditches

The occasional management of ditches to clear accumulated silt that has blocked the field drain outfalls will be continued. During the winter of 1987–8, the Severn Trent Water Authority (now part of the National Rivers Authority) carried out routine maintenance work on Squitter Brook. In addition to removing accumulated silt, the Authority agreed to add several small pools on the stream, pull back the banks and raise the water level at one point with a small, stone weir. This has improved the conservation value of the brook considerably, varying the flow and depth and allowing more light to the water.

The farmer and farm manager agreed with the adviser that one of the field margins to be left unploughed (see field margin management) should be sited alongside the species-rich Deerfield Ditch, where it would help protect the wildlife in the ditch from field operations.

Stone walls

Of the few stone walls still forming field boundaries on the farm, those in visually prominent places in the village of Kemerton, or beside tracks, are repaired as and when necessary by farm staff. Elsewhere, on the higher ground, the stone walls have no agricultural value and have been allowed to crumble. Cost and the low priority that they receive mean that their restoration is not a practical option in the foreseeable future.

Summary

A summary of the options proposed by the adviser for boundary management on the case study farm is shown below. It includes a record of which options were selected for implementation, which rejected and which held over pending further discussion. The numbers indicate the positions of the selected options on the farm management plan (Figure 6.6).

▶ New hedge to divide large, hillside field – No

▶ New hedge along roadside – Agreed (1)

▶ New hedge on hillside beside footpath – Agreed (2)

▶ New hedge between garden and field – Agreed (3)

▶ Low internal hedges to be trimmed less frequently – Perhaps

▶ Parish boundary hedge to be allowed to grow taller – Agreed (4)

▶ One hedge to be coppiced next winter – Agreed (5)

▶ Practice of allowing certain hedges to grow up and laying a stretch each winter to continue – Agreed (6)

▶ New hedgerow trees in low hedges – Agreed (7)

▶ Long-term programme of pollarding willows to continue – Agreed (8)

▶ Uncultivated margin of 1–1.5 m to be left round all arable fields – Partly agreed (9)

▶ Conservation headlands to continue – Agreed

▶ Management of existing grass margins to continue – Agreed

▶ Two new grass margins to be established by sowing along weedy field margins – Agreed (10)

▶ Introduction of wild flowers into existing margins – Perhaps

▶ Arable weed margins to continue – Agreed

▶ Existing infrequent management to continue – Agreed

▶ Grass margin buffer zone to be established alongside species-rich ditch bank – Agreed (9)

Chapter 7
IMPLEMENTING AND MONITORING A MANAGEMENT PLAN

Once you have decided on the most appropriate forms of management for the boundaries on your land, it is good practice to incorporate these into a formal management plan. This will be especially valuable if others will be involved in carrying out the work, or where responsibility for the area concerned is shared or likely to change hands from time to time.

It is best to keep the plan simple, so that it is easily understood by others, but it should provide enough detail to allow you to monitor progress and check that particular tasks have been carried out. It needs to include not only your chosen options, but how and when these are to be implemented. For example, rather than just stating that some hedgerow saplings should be left to grow into trees, you should say when, where and, if necessary, how this will be done, which species are preferred and the approximate or minimum spacing: 'summer, year 1, select and tag five to six sturdy saplings of small tree species (e.g. hawthorn, field maple) in the hedge along the footpath, at a minimum spacing of 15 metres apart, to be retained as amenity trees'.

The plan should include information on the care of new features for the first five years, and indicate any longer-term management required. The proposed timing and sequence of operations in the short and long term can be recorded in diary form as a reminder.

Except for small-scale projects or minor adjustments to routine tasks, you may also need to include on the plan an indication of any additional labour, materials and equipment required, and the approximate costs. Before the plan is implemented, investigate whether any legal restrictions apply, and follow up possible sources of grant aid. Bear in mind any safety precautions that must be taken, and consider whether you need to arrange training for those who will carry out the work.

When you have completed the plan, it is a good idea to keep a copy of it, or a summary, in a prominent position, for example pinned to the office wall, where it can readily be consulted once work is under way.

This chapter summarises the points that you should check in relation to labour and equipment requirements, costs, grants, safety, training and legislation. It also discusses ways of informing others of your aims, both to ensure that the management plan is implemented as you wish and to avoid upsetting neighbouring landowners and residents. Finally, it emphasises the importance of monitoring progress, so that you can have the satisfaction of noting any improvements that occur as a result of your efforts and can modify your plan if things do not go as expected.

7.1 Labour

Many of the options described for maintaining, enhancing and creating boundary habitats involve relatively small adjustments to routine management practices, which can usually be accommodated within existing resources of labour and machinery. Examples include the accurate placing of fertilisers and pesticides so that they do not drift over boundary areas, the tagging of hedgerow saplings and the establishment of arable weed margins. Other options, although part of routine maintenance, may require the help of contractors, either because special skills or expensive equipment are involved, or because they coincide with busy times of year. Examples include hedge trimming and ditching. A few tasks, which are infrequently undertaken and very time consuming, such as hedge laying and stone walling, will almost certainly require additional professional help. Such tasks, which are now often too costly to be undertaken as part of routine land management, but which are of considerable landscape and wildlife value, are among those most likely to qualify for conservation grants. This justifies a thorough investigation of possible sources of help before work begins (see Section 7.4 on grants). A guide to the labour requirements for a range of boundary management tasks is shown in Table 7.1.

7.2 Machinery and equipment

Even for tasks that are to be undertaken by contractors, it is useful to do some research on the type of equipment available, so that you can specify what is most appropriate for your particular needs. Check that machinery is manoeuvrable enough to negotiate any narrow paths and lanes to the site,

Table 7.1 Guide to labour requirements for boundary habitat management

	Approximate work rate*	Time of year	Frequency
Hedge trimming (flail)	5 km per day	Aug–Feb (preferably Dec–Feb)	Every 2–3 years (preferably not annually)
Hedge laying	5–20 m per day	Nov–Mar	Every 10–20 years
Hedge coppicing	100 m per day (two people with a chain saw) 13 m per hour (contractor with a tractor-mounted saw, a driver and two other people)	Late autumn, winter	
Hedge planting	20–40 m per day	Oct–Mar	
Dry stone walling	2–5 m per day		
Pollarding	Two or three trees per day		Every 20–40 years

*This is intended to provide only a rough guide; the rate is likely to vary considerably with such factors as the skill of the person(s) involved, the condition of a hedge, whether a stone wall is being repaired or built from scratch.

and that it has sufficient capacity to cope with the expected workload at the required time of year.

In the case of hedge cutting, there are four principal types of tool that can be used: cutter bars, flail cutters, circular saws and hand tools. The choice should be determined mainly by the condition of the hedge. Tractor-mounted cutter bars can be used on hedges trimmed every year, but are not designed to deal with heavy growth or branches more than 5 centimetres in diameter. Flail cutters are the machines most widely used for trimming back the growth on hedges, verges and ditches; they chop the vegetation to a fine mulch so that little clearing up is needed afterwards. A light-duty flail will be able to cut material up to 2.5 centimetres in diameter. Heavy-duty flails can cut quite stiff wood up to 10 centimetres in diameter, but the resulting slashed and split stems look very unsightly and leave the shrubs susceptible to fungal disease and die-back. Hedges that have not been regularly trimmed, with stems between 5 and 15 centimetres in diameter, should be cut with circular or shape saws, which can be either tractor mounted or trailed. The resulting trimmings need to be collected and left in heaps to rot down out of the way of paths, or burnt if there is a risk of arson. Larger stems (more than 15–20 centimetres in diameter) should be cut with power chain saws before mechanical trimming. Traditional hand tools, such as lightweight axes, billhooks, hedge slashers and hedge saws, are still used where access for machinery is difficult, for cutting out elder, climbers and bramble and to trim round hedgerow saplings that are to be left to grow.

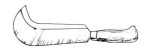

A similar range of equipment can be used on grass verges. Cutter bars, which leave the cut vegetation in one piece, are better than flail cutters, which produce a fine mulch, if the aim is to rake up and remove the cuttings to encourage a fine sward. Forage harvesters have been used as an alternative way of collecting the cuttings and directing them away from the verge onto an adjoining field. Flail cutters can be used to prevent the encroachment of scrub on grass verges, provided that the cutting is done every two or three years so that the growth is not too woody. For lineside vegetation, specially adapted flailing machines which run along the railway track on wheels are available.

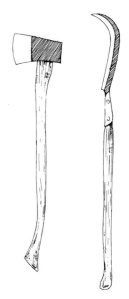

Traditional crafts, such as hedge laying and stone walling, have their own range of hand tools, often with distinctive regional variations. The hedge layer's basic equipment usually includes: a billhook for general cutting and trimming; an axe for cutting the pleachers; a slasher or long-handled hook for preliminary cutting out of excess brushwood and freeing the tops of tangled pleachers before laying them; a bow saw or chain saw for dealing with thicker stems; a mallet for knocking in stakes; and a sharpening stone and flat file for maintaining a sharp cutting edge on the tools. Stone walls can be built, or at least repaired, without any tools, but useful equipment includes: one or more walling hammers for trimming and splitting stones; a sledge hammer or club hammer for breaking up bigger stones; a heavy-duty spade for digging foundations, cleaning round the base of existing walls or building stone-faced banks; a pick or pick-ended mallet to supplement the spade and dig up field stones; a batter frame as a template for walls; a plumb bob and line; a spirit level; and a steel tape measure. Buckets, sledges and wheelbarrows are useful for carting the stones.

As well as the equipment needed to manage a boundary itself, some thought needs to be given to any equipment used to apply fertilisers and sprays to the adjoining land, since this can have an important influence on the value of the habitat. Pneumatic, full-width fertiliser distributors can be manouevred to restrict fertiliser application to the cropped area, but oscillating spout or spinning disc distributors are designed to work with an overlapping pattern to give an even application of fertiliser overall (Figure 7.1a). Unless the pattern of spread is altered for the boundary run, some fertiliser is applied to the boundary area, which is both harmful to wildlife and a waste of money. Several manufacturers now offer inexpensive kits to modify the spreading pattern for the boundary run. Some machines incorporate a mechanism that allows the whole distributor to be tilted to offset the pattern of spread (Figure 7.1b). In the case of crop sprayers, which have nozzles designed to produce an overlapping pattern of spray (Figure 7.2a), application beyond the crop edge can be prevented by turning the end nozzle tip at an angle of 45° to the line of travel (Figure 7.2b), by fitting an off-centre fan tip if bayonet fitting nozzles are used (Figure 7.2c) or by shutting off the end nozzles completely (Figure 7.2d).

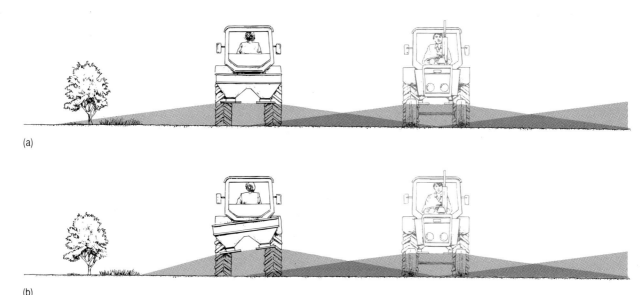

(a)

(b)

Figure 7.1   Preventing the application of fertiliser to field boundaries

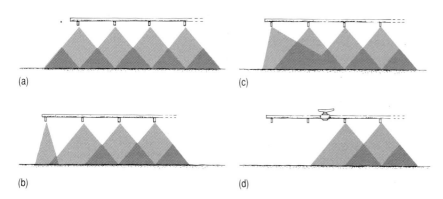

(a)

(b)

(c)

(d)

Figure 7.2   Preventing the application of sprays to field boundaries

Where ditch vegetation has to be controlled, hand cutting with scythes and shears is still widely practised, especially where access for machinery is difficult or where herbicides cannot be used. Mechanised cutting involves the use of weed cutting buckets mounted on hydraulic or dragline excavators. This may be supplemented by the use of tractor-mounted flail mowers to cut the vegetation on the bank above the water line. Cutting machines are now being developed that straddle water courses, to avoid some of the problems of access faced by conventional machines. Only a limited range of herbicides have been approved for weed control on or near water; professional advice on appropriate chemicals and methods of application should be sought from organisations such as the National Rivers Authority and the Agricultural Development and Advisory Service (see the *Helpful Organisations* supplementary booklet in the foundation module).

7.3 Costing boundary management

Small projects, and those that involve only minor adjustments to existing management regimes, are unlikely to need a detailed assessment of the costs involved. If you are convinced of the benefits of the scheme and feel that any costs can readily be absorbed, this will probably provide sufficient justification to go ahead. For larger schemes, and work undertaken on behalf of others, a clear indication of costs and benefits is likely to be required. How comprehensive the costings need to be will depend on your purpose.

The most important costing component will probably be an estimate of variable costs, covering any additional materials, equipment or labour (casual or contract) needed specifically for boundary management. These costs vary with the area or length of boundary involved. Fixed costs, such as the payment of permanent labour and depreciation on equipment that you already own, should not be included, unless the extra work involved in boundary habitat management means that a threshold is reached above which more permanent labour or an extra machine is needed.

Where detailed costings are required, you may need to include opportunity costs and discounted costs. Opportunity costs are the equivalent of the income you might have received if the resources of labour, land and capital involved in the management of a boundary area had been used for other, more profitable, purposes, for example if an arable weed margin had been planted with winter wheat. The opportunity costs should take into account any reductions in expenditure as a result of boundary management, for example reduced inputs of seed, fertiliser and sprays in a field as a result of the establishment of a margin. For long-term and expensive projects, discounted costs are sometimes used in weighing up the relative merits of different forms of investment, for example in comparing the cost of installing new fencing with that of planting a hedge. Discounted costs are the costs of managing a new feature in the foreseeable future discounted to the present day. On this basis, a low initial cost, high maintenance cost option may be preferable to a high initial cost, low maintenance cost alternative. A formal discounted cost appraisal may rarely be needed, but it is worth bearing in mind likely differences in long-term maintenance costs when choosing among options.

Offset against costs can be any grants to which you are entitled, any income that you expect to receive from the boundary feature, for example from charging for leisure pursuits, and the less tangible benefits, such as more pleasant surroundings. Although intangible benefits cannot be measured, it is a good idea in any formal costings exercise to describe their value in words, both to ensure that they are fully taken into account and as a reminder to others that they exist.

7.4   Grants and other forms of help

It is important to check possible sources of grant aid before work begins, since grants cannot usually be approved retrospectively. If your land is in a designated area, such as a National Park or an Environmentally Sensitive Area, grant aid will almost certainly be available for some forms of boundary habitat management. In these cases, the grants are provided by statutory authorities such as English Nature, the Countryside Commission, the Agricultural Development and Advisory Service and their equivalents in Scotland and Wales.

Outside designated areas, local authorities and occasionally local groups such as the county wildlife trusts sometimes give grants towards boundary management that benefits conservation. Alternatively, wildlife trusts may take on the management of small areas that are particularly valuable for wildlife, although they may prefer a management agreement under which the landowner is recompensed for managing the site in a particular way. Groups such as the British Trust for Conservation Volunteers and the Scottish Conservation Projects Trust can provide labour for conservation tasks, in some cases for a minimal charge to cover expenses. Local schools and colleges may be interested in helping with species identification and monitoring, especially if this can be linked with project work.

Expert practical conservation advice can be obtained from organisations such as the Agricultural Development and Advisory Service and the Farming and Wildlife Advisory Group, and the initial consultation is usually free of charge. Expert advice on species identification can be obtained from local natural history societies, wildlife groups and English Nature, among others. Details of all these and other sources of help and advice are given in the *Helpful Organisations* supplementary booklet in the foundation module.

7.5   Safety and training

In implementing your management plan, safety considerations always need to be kept in mind. This is especially the case when young or inexperienced workers are helping out. The safety of visitors and members of the general public should be considered, particularly where boundaries are used for leisure pursuits or there is public access. A first-aid kit should always be on hand for dealing with minor accidents.

Clothing

Overalls, or close-fitting but comfortable work clothes, should be supplemented with protective clothing when necessary. Heavy work boots with a good grip to the soles are needed for jobs like hedge laying; the boots should be reinforced with steel toe caps when heavy weights are involved, as in the case of stone walling. Stout leather gloves should be worn for tasks such as hedge laying, where infection can occur as a result of thorn wounds and cuts. Some form of protection from crushed fingers should also be worn for stone walling, although many professional stone wallers find it difficult to work in gloves, and instead may wear mittens or protective patches improvised from rubber tubing. Eye protection is needed for jobs such as cutting wood and chipping and splitting stone.

Hand-held tools, such as hammers and axes, should be of good quality, with the heads made of well-tempered steel. All equipment and machinery should be checked to make sure it is in good order before work begins, for example that blades are sharp and that the flails on hedge cutters are firmly fixed. The cabs of tractors being used with flails and circular saws should be fitted with wire mesh screens to protect drivers from flying debris.

Equipment and machinery

Before machinery is used, an area should be checked for potentially hazardous obstructions, such as overhead cables, pieces of wire and metal and large stones. The position of immovable obstacles that might not be noticed should be clearly marked. The slope of the ground and any ruts or irregularities that might affect tractor stability should be taken into account. Tractor stability is likely to be reduced when offset tools, such as hedge trimmers, are being used.

Obstructions

Workers should keep a safe distance apart when using sharp-edged tools or machines. Chain saws in particular should be used with extreme care; full protective clothing should be worn and the person carrying out the work should not be left alone in case of an accident. Those working with stone should take care when lifting and ensure that falling or rolling stones do not endanger others.

Work practices

Extra care is needed when machinery is being used on a roadside. Advance warning signs should be used to alert motorists, and the road surface cleared afterwards of debris, such as hedge trimmings.

When chemicals are being used, the manufacturer's instructions should be carefully followed and protective clothing worn if necessary. Legislation concerning records of the chemicals used, checks on spraying equipment and training required for those handling the sprays should be complied with and the MAFF code of practice followed.

Chemicals

Training is essential for anyone carrying out work on lineside vegetation and for most pesticide users. Training courses for a variety of practical land management tasks, such as coppicing, hedge trimming, hedge laying, stone walling, chain saw use and spraying, are offered by organisations such as the Agricultural Training Board, the British Trust for Conservation Volunteers and the Scottish Conservation Projects Trust (see the *Helpful Organisations* supplementary booklet in the foundation module).

Training

7.6   Legislation and regulations

Before work begins, ensure that you are familiar with any legislative requirements that apply. Details of those with particular relevance for conservation are given in the *Legislation and Regulations* supplementary booklet in the foundation module. General regulations likely to affect boundary habitat management include those covering the safety of people at work (for example the Health and Safety at Work etc. Act 1974) and the use of pesticides (for example the Food and Environment Protection Act 1985).

Legislation governing species and habitat protection, which may affect some boundary areas, includes the Wildlife and Countryside Act 1981 and the Badgers Act 1973. Legislation that applies in particular to boundaries includes the Rights of Way Act 1990 (covering maintenance responsibilities for footpaths and bridleways) and the Environmental Protection Act 1990 (covering, amongst other things, pollution control for water courses).

It is particularly important at the present time, when public interest in environmental matters and membership of the European Community is forcing the pace of regulatory change, to keep up to date with any new control measures being introduced, for example by reading trade journals. In some cases, restrictions are linked with grants by means of management agreements. Thus, proposals for the introduction of Hedgerow Management Orders, which are currently being discussed as a way of protecting the most highly valued hedges, include recommendations for payments towards the cost of their maintenance.

7.7   Informing others

Where your plan is to be implemented by others, or where operations carried out by others are likely to affect your plan, it is essential that those involved understand clearly what you require to be done (or avoided) and why. Particularly valuable stretches of boundary habitat should be clearly marked if they require special management or protection. Several counties have designated their more interesting stretches of roadside verge as nature reserves. In Bedfordshire, these are marked with posts, set well back from the road so that they do not distract drivers, each bearing a simple notice that informs contractors that the verge is only to be cut once a year in the autumn. The notice also serves to inform curious passers-by of the existence of such reserves, and indicates how they can find out more about them. The position of all the roadside reserves is marked on a map lodged with the planning department, so that whenever organisations seek permission for a new development, or to carry out work on underground services, such as gas and water, on or near a verge, they can be advised what needs to be done to avoid or minimise damage.

While maintenance or development work is in progress, areas of high conservation value can be marked off with coloured tape to avoid accidental damage, for example by the dumping of materials and soil.

Since by their very nature boundary habitats often separate land under different ownership, there will be many circumstances where the work you do on a boundary will affect your neighbours. When this is the case, it makes sense to ensure that they are kept informed of what you are doing and why, especially where a boundary adjoins public or residential areas. Highly visible forms of management, such as the coppicing of a hedge or the felling of mature trees, can meet with hostile reactions unless their purpose is explained. Depending on the scale of the project, this information can be provided by means of a noticeboard on the boundary, leaflets to local houses, a note in the parish magazine or local newspaper, or a public meeting. As experience with the Hampshire Railway Lineside Vegetation Project (see Box 4.1) has shown, hostility usually disappears once people are informed about what is proposed. They may even come forward with constructive suggestions and offers of help.

7.8   Monitoring progress

The management of boundary features is a cyclical process, although the cycle may vary in length from a few weeks for path mowing to scores of years in the case of hedgerow trees. Management of the land adjoining a boundary may change, and this in turn may affect management of the boundary. As a result, a management plan can never be said to be truly completed, and should be regularly updated as circumstances change. Minor adjustments to the plan can be made as needed, and the plan reassessed and overhauled about once every five years.

Updating the plan will be made easier by regular checks on progress, to ensure that the desired effect is being achieved and that no unforeseen problems have occurred. Monitoring can take the form of a regular site visit to check, for example, that new hedgerow shrubs are putting on growth and not being choked by weeds or killed by drought, or that markers to identify valuable sites remain in place. Photographs taken at regular intervals from the same spot at the same time of year can provide a useful record of progress, as can walking the same stretch of boundary regularly and recording any changes in the number and type of different flowers, butterflies or birds seen. Bear in mind that natural fluctuations in wildlife numbers can occur from year to year, for example as a result of particularly harsh winters, and that some changes are very gradual, in which case it may be some years before improved management has a noticeable effect.

In many ways, some of the simplest and least costly options, such as protecting boundaries from disturbance and trimming hedges a little less often, can bring the most rewarding results. Whether changes are immediately obvious or occur gradually, a management plan will provide an evolving record of your achievements. By maintaining, enhancing and adding to the boundary features that already exist, you will have the satisfaction not only of increasing the interest of your own surroundings but of adding to the great pleasure that the British countryside gives to many others.

Appendix I
FURTHER READING

British Trust for Conservation Volunteers (1975) *Hedging*. British Trust for Conservation Volunteers, Wallingford, Oxfordshire

British Trust for Conservation Volunteers (1986) *Drystone Walling*. British Trust for Conservation Volunteers, Wallingford, Oxfordshire

Cobham Resource Consultants (1984) *Landscape and Wildlife Conservation on Farms*. The Countryside Commission for Scotland, East of Scotland College of Agriculture and the Nature Conservancy Council

Cobham Resource Consultants (1985) *Landscape and Wildlife Conservation on Farms: A Further Study*. The Countryside Commission for Scotland, East of Scotland College of Agriculture and the Nature Conservancy Council

Countryside Commission (1987) *Landscape Assessment: A Countryside Commission Approach*, CCD 18. Countryside Commission, Cheltenham, Gloucestershire

Countryside Commission (1990) *Changes in Landscape Features in England and Wales 1947–1985*, CCD 44. Countryside Commission, Cheltenham, Gloucestershire

Darlington, A (1981) *Ecology of Walls*. Heinemann, London

Department of Transport, Scottish Development Department, Welsh Office (1990) *Transport Statistics Great Britain 1979–1989*. HMSO, London

Dowdeswell, W H (1987) *Hedgerows and Verges*. Allen & Unwin, London

Harris, S, Jefferies, D, Cresswell, W (1988) *Problems with Badgers?* Royal Society for the Prevention of Cruelty to Animals, Horsham, Sussex

Hoskins, W G (1955) *The Making of the English Landscape*. Hodder & Stoughton, London

Hoskins, W G (1973) *English Landscapes*. British Broadcasting Corporation, London

Lee, B (1985) *Guide to Fields, Farms and Hedgerows*. The Crowood Press, Ramsbury, Wiltshire

Muir, R, Muir, N (1987) *Hedgerows: Their History and Wildlife*. Michael Joseph, London

Nature Conservancy Council (1989) *Nature Conservation and the Management of Drainage Channels*. Nature Conservancy Council, Peterborough

Pollard, E, Hooper, M D, Moore, N W (1974) *Hedges*. Collins, London

Rackham, O (1986) *The History of the Countryside*. J M Dent, London

Sturrock, F, Cathie, J (1980) *Farm Modernisation and the Countryside*. University of Cambridge, Department of Land Economy

Taylor, C (1983) *Village and Farmstead*. George Philip, London

Way, J M (ed.) (1983) *Management of Vegetation*. British Crop Protection Council, Thornton Heath, Surrey

Way, J M, Greig-Smith, P W (eds) (1987) *Field Margins*. British Crop Protection Council, Thornton Heath, Surrey

Leaflets on practical topics

A number of organisations publish leaflets on specific aspects of boundary habitat management, such as those listed here.

Agricultural Development and Advisory Service (1986) *Hedgerows P3027*. MAFF Publications, Alnwick, Northumberland

Agricultural Development and Advisory Service (1988) *Shelter Belts in the Uplands P3154*. MAFF Publications, Alnwick, Northumberland

Agricultural Development and Advisory Service (1989) *Trees for Shelter on the Lowland Farm P3187*. MAFF Publications, Alnwick, Northumberland

Bomford and Evershed Ltd (undated) *Hedges and Hedgerow Management*. Bomford and Evershed, Evesham, Worcestershire

Countryside Conservation Handbook (1980) *Leaflet 7 Hedge Management*. Countryside Commission, Cheltenham, Gloucestershire

Countryside Conservation Handbook (1982) *Leaflet 10 Game Habitats on the Farm*. Countryside Commission, Cheltenham, Gloucestershire

Countryside Conservation Handbook (1982) *Leaflet 12 Conserving Farm Ditches and Watercourses*. Countryside Commission, Cheltenham, Gloucestershire

Farming and Wildlife Advisory Group (1985) *Butterflies on the Farm. FWAG Information 16*. Farming and Wildlife Advisory Group, Stoneleigh, Warwickshire

Forestry Commission (1990) *The Establishment of Trees in Hedgerows*. Forestry Commission, Farnham, Surrey

Ministry of Agriculture, Fisheries and Food (1986) *Field Drainage and Conservation Booklet 2522*. MAFF Publications, Alnwick, Northumberland

Nature Conservancy Council (1979) *Hedges and Shelterbelts*. Nature Conservancy Council, Peterborough

Nature Conservancy Council (1982) *The Conservation of Farm Ponds and Ditches*. Nature Conservancy Council, Peterborough

Schering Agriculture (1988) *Farm Conservation Guide*. Schering Agriculture, Nottingham

Schering Agriculture (1989) *The Management of Cereal Field Margins*. Schering Agriculture, Nottingham

Suffolk and Norfolk County Councils (undated) *Breckland Scots Pine Lines and Belts*. Suffolk County Council, Ipswich and Norfolk County Council, Norwich

The Game Conservancy (undated) *Guidelines for the Management of Field Margins (Conservation Headlands and Field Boundaries)*. The Game Conservancy Trust, Fordingbridge, Hampshire

Wiltshire Farming and Wildlife Advisory Group (undated) *Making the Most of Your Hedges* (series of leaflets). Wiltshire Farming and Wildlife Advisory Group, Shrewsbury. (Many county Farming and Wildlife Advisory Groups have similar leaflets)

Wildlife identification

There are numerous books on the market to help with the identification of plants and animals in Great Britain. The following list gives some examples.

Collins Field Guides

Arnold, E N, Burton, J A, Overden, D W (1978) *The Reptiles and Amphibians of Britain and Europe*

Bang, P, Dahlstrom, P (1984) *Animal Tracks and Signs*

Barrett, J, Younge, C M (1958) *The Sea Shore*

Chinery, M (1976) *The Insects of Britain and Northern Europe*

Fitter, R, Fitter, A, Blamey, M (1978) *The Wild Flowers of Britain and Northern Europe*

Kerney, M P, Cameron, R A D, Riley, G (1979) *The Land Snails of Britain and North West Europe*

Large, M, Hora, F B (1965) *Mushrooms and Toadstools*

Mitchell, A (1978) *The Trees of Britain and Northern Europe*

Peterson, R, Mountfort, G, Hallam, P A D (1974) *The Birds of Britain and Europe*

Mitchell Beazley Pocket Guides

Hayman, P (1979) *Birds*

Moore, P D (1980) *Wild Flowers*

Rushford, K (1980) *Trees*

Pan Books

Hammond, N, Everet, M (1980) *Birds of Britain and Europe*

Phillips, R (1977) *Wild Flowers of Britain*

Phillips, R (1978) *Trees in Britain*

Phillips, R (1980) *Grasses, Ferns, Mosses and Lichens of Great Britain and Ireland*

Phillips, R (1981) *Mushrooms*

Appendix II
GLOSSARY

Acid Term used to describe soils with a low pH, such as peat.

Acts of Enclosure Acts passed by parliament, mainly between 1750 and 1850, permitting the enclosure of open fields and other commonly held land. Enclosure commissioners were appointed in each parish to oversee the fair redistribution of land; once land was reallocated the new landowner had to mark its outer boundary with a hedge or wall within a year.

Algal bloom Dense growth of algal weed, which develops in water that has been enriched by fertiliser.

Alkaline Term used to describe soils of a *calcareous* nature with a high pH, such as those on chalk or limestone.

Alluvial Term used to describe material deposited by streams and rivers. .

Ancient countryside Districts in which the fields, woods and roads date predominantly from before AD 1700.

Anglo-Saxon charter Legal conveyancing document for a piece of land, often describing the land's boundaries, dating mainly from the period AD 600–1080.

Annual Plant that completes its life-cycle, from seed germination to seed production followed by death, within a single season.

Area of Outstanding Natural Beauty (AONB) Area designated by the Countryside Commission as of particular landscape beauty, smaller in size than a *National Park* and without a managing authority.

Assart Area cleared for farming, usually from woodland but sometimes from moor or fen.

Batter frame Wooden or metal frame used as a guide to the correct batter and to the heights of throughs, topstones, etc. when building a wall or hedge. Also known as a pattern (South-West), template or wall gauge (Cotswolds) or walling or dyke frame (Scotland).

Baulk Uncultivated grassy strip or earth bank, which separated groups of cultivated strips in the *open field agriculture* system.

Berm Ledge.

Black Death Outbreaks of bubonic plague, which decimated European and British populations in the mid fourteenth century.

Break crop Crop grown to break a succession of cereal crops, for example oilseed rape, peas or beans, to help avoid a build-up of cereal pests and weed and disease problems.

Bronze Age Period of history characterised, among other things, by the use of bronze; 2400–750 BC.

Calcareous Made of, or containing, calcium carbonate and therefore *alkaline*.

Calcicole Plant that thrives on *calcareous* or *alkaline* soils, such as those on chalk or limestone.

Callus Wound tissue, for example on the trunk of a *pollarded* tree.

Cess path On a railway, the inspection path alongside the track.

Conservation headland Field *headland* on which the use of pesticides and/or fertiliser is reduced or eliminated to encourage plant and animal wildlife.

Cope stone Stone used to top or cap a wall (also known as a *top stone*).

Coping Layer of *cope stones* that caps a wall.

Coppicing Cutting shrubs or trees close to the ground to allow new shoots to grow from the stumps or *stools*, on a rotational basis.

Cripple hole Opening at the base of a wall designed to allow sheep but not cattle through.

Culvert Arched channel for carrying water under a road or railway.

Deciduous Term used to describe a shrub or tree that sheds its leaves over winter.

Diameter at breast height (dbh) Diameter of a tree measured 1.3 metres above ground level (a standard forestry measure).

Dike See *dyke*.

Drove road Old, generally grassy, track used to move cattle and other livestock between pastures and to market.

Dutch elm disease Disease of elms caused by the fungus *Ceratostomella ulmi* and spread by the elm bark beetle (*Scolytus scolytus*); estimated to have killed at least 40% of all elms in Britain in the late 1960s.

Dyke Stone wall in Scotland; ditch in Norfolk.

Edder Pliable, woody stem woven along the top of the stakes in a laid hedge to keep the *pleachers* in place (also known as a binder, ether, heather, heathering).

Enclosure Award Deed showing entitlement to land redistributed by the enclosure commissioners as a result of the *Acts of Enclosure.*

Environmentally Sensitive Area (ESA) Area designated by the Ministry of Agriculture, Fisheries and Food, the Department of Agriculture for Scotland or the Welsh Office of Agricultural Development as a place in which it is deemed desirable to conserve and enhance the natural beauty, conserve the flora or fauna or geological or physiographical features, or protect buildings or other objects of archaeological, architectural or historical interest.

Face stone Stone whose outer surface forms part of the face of a wall (also known as a builder).

Fallowed headland Field *headland* that is cultivated but not cropped, to encourage the growth of *annual* arable weeds.

Fauna Animal population present at a particular place or time.

Fillings Small, irregular stones placed between the two faces of a wall to pack the space between them (also known as middlings).

Flail strip On a railway, a narrow strip of vegetation alongside the *permanent way*, which is regularly flailed (or sprayed with herbicide) for operational reasons.

Flora Plant population present at a particular place or time.

Food chain Series of organisms, beginning with plants and ending with the larger animal predators, which are dependent upon others in the series for food.

Food web Interdependent *food chains* in a community.

Foundation stone Stone at the base of a wall (also known as a footing).

Green lane Old lane or track that has not been adopted and surfaced as part of the modern road system.

Headland Area at the ends of a field in which cultivation machinery is turned round and which has to be cultivated separately from the main body of the field.

Hydroseeding Process by which seeds are applied to an embankment or cutting by spraying with a mixture of water and a bulky, organic material, such as pulverised wood, containing the seed. The bulky material helps to trap the seeds in place on sloping ground.

Insectivorous Term used to describe an organism that feeds on insects.

Layering Practice of bringing part of a shoot into contact with soil to encourage it to take root and form a new plant at that point. (The term is sometimes used to mean *laying*.)

Laying Management of a hedge by cutting part way through the stems of shrubs and trees and bending the cut stems sideways to form a barrier (also known as *layering*, pleaching and plashing).

Lias Sedimentary rock deposited at the beginning of the Jurassic period (140–195 million years ago).

Lichen Small plant formed by the symbiotic association of a fungus and an alga.

Lodging Collapse of the stems of a crop so that they lie flat on the ground.

Loess Deposits of wind-transported silt.

Lynchet Terrace or ridge formed in prehistoric or *medieval* times by ploughing a hillside.

Mearestone Stone that marks a boundary.

Medieval Period of history from AD 1066 to AD 1536, the Middle Ages.

Mulch Material used to cover the soil surface, conserving soil water and discouraging weed growth.

National Park Area designated by the Countryside Commission by reason of natural beauty and opportunities afforded to the public for open air recreation, which has its own planning and management arrangements.

National Scenic Area (NSA) Area designated by the Countryside Commission for Scotland as of particular landscape beauty.

Nematode Roundworm or threadworm belonging to the Class Nematoda, some of which are parasitic on plants and animals.

Neutral Term used to describe soils that are neither *acid* or *alkaline*, with a pH of approximately 7.

Oolitic limestone Sedimentary rock derived from a build-up of small, *calcareous* spheres, deposited during the Jurassic period (140–195 million years ago).

Open field agriculture System of agriculture particularly common in the Midlands in the twelfth and thirteenth centuries, in which individual farmers cultivated strips of land aggregated into large communally farmed open fields (also known as strip cultivation).

Perennial Plant that continues its growth from year to year.

Permanent way On a railway, the area that includes the track and track inspection walkways and which is kept free of vegetation.

Permissive path Path that is used by the public by permission of the owner of the land over which the path runs, with the intention that it does not become a public right of way. The landowner may erect a notice to this effect and perhaps close the path once a year.

Planned countryside Districts in which the fields, woods and roads were laid out predominantly in the eighteenth or nineteenth century.

Pleacher Live, woody stem cut part way through and laid, to form a stock barrier (also known as a plasher, plesher, pletcher, pusher, sear or stolling).

Pollarded Term used to describe a tree the trunk of which has been cut 2–4 metres above ground level and then allowed to grow again to produce a crop of branches.

Rhine Ditch in Somerset.

Rhyne See rhine.

Ridges and furrows Regular undulations still visible in some old pasture, usually an indication of *medieval* strip cultivation.

Ridgeway Ancient, long-distance road or track running along hill tops.

Rods Crop of young, unbranched stems obtained in the first year or so after willow is pollarded or coppiced; colloquially known as *withies* and distinguished from poles by their size (rods are normally narrow stems less than 2 centimetres in diameter, whereas poles are older stems about 4 centimetres across).

Roman settlement In England, period of history from AD 40 to AD 410.

Saxon In England, period of history from AD 410 to AD 1066.

Shelter-belt Band of shrubs and/or trees planted to provide protection from the weather.

Site of Special Scientific Interest (SSSI) Area designated by the Nature Conservancy Council as being of special interest because of its flora, fauna, geology or physiography.

Smoot Small hole in the base of a wall to allow water to drain through and rabbits to pass through.

Squeezer Stile that consists of a gap just wide enough to allow people through but too narrow for sheep.

Stool Stump or cut base of a shrub or tree from which new shoots grow. See also *coppicing*.

Succession Replacement of one type of community by another, shown by progressive changes in vegetation and animal life.

Sucker Woody shoot arising from an underground stem or root, or shoot arising from the rootstock of a worked (grafted or budded) plant.

Through stone Large stone placed across the width of a wall to link the two sides together (also known as a bonder).

Top stone See *cope stone*.

Towpath Beside a canal, path once used by the horses that towed canal barges.

Transplant Plant that has been transplanted one or more times in a nursery to develop a better root:shoot ratio than would be the case if the plant was allowed to grow undisturbed. Usually refers to a two- to three-year-old shrub or a tree 0.2–0.45 metre in height.

Tree Preservation Order (TPO) Local planning authority designation which can be placed on a tree, a group of trees or woodland to help conserve the amenity of an area.

Turnpike road Road on which there were (or are) toll-gates.

Whip Young tree 0.6–2.1 metres high, larger than a *transplant* but smaller than a standard tree.

Withies Somerset term used to describe basket willows and basket willow beds; sometimes used to describe the crop of young, flexible willow branches obtained after pollarding and used for basket making (also called *rods*, canes or wands).

Xerophyte Drought-tolerant plant.

Appendix III
SCIENTIFIC NAMES FOR WILDLIFE SPECIES

Plants

Adders tongue fern *Ophioglossum vulgatum*

Adders-tongue spearwort *Ranunculus ophioglossifolius*

Agrimony *Agrimonia eupatoria*

Alder *Alnus glutinosa*

Arrowhead *Sagittaria sagittifolia*

Ash *Fraxinus excelsior*

Aspen *Populus tremula*

Barberry *Berberis vulgaris*

Barren brome *Bromus sterilis*

Beech *Fagus sylvatica*

Bents *Agrostis* spp

Bindweed,
    Black *Polygonum convolvulus*
    Field *Convolvulus arvensis*
    Hedge *Calystegia sepium*

Birch,
    Downy *Betula pubescens*
    Silver *Betula pendula*

Birds-foot-trefoil *Lotus corniculatus*

Bittersweet *Solanum dulcamara*

Black alder (see Alder)

Black-grass *Alopecurus myosuroides*

Black horehound *Ballota nigra*

Black medick *Medicago lupulina*

Black nightshade *Solanum nigrum*

Black spleenwort *Asplenium adiantum-nigrum*

Blackthorn *Prunus spinosa*

Blanket weed *Cladophora* spp

Bluebell *Hyacinthoides non-scripta*

Box *Buxus sempervirens*

Bramble *Rubus fruticosus*

Briars (see Rose; Sweet briar)

Broad-leaved dock *Rumex obtusifolius*

Broad-leaved pondweed *Potamogeton natans*

Brooklime *Veronica beccabunga*

Broom *Cytisus scoparius*

Brown top bent (Common bent) *Agrostis tenuis*

Bryony,
    Black *Tamus communis*
    White *Bryonia dioica*

Buckthorn,
    Alder *Frangula alnus*
    Purging *Rhamnus catharticus*

Bullace (see Plum)

Burnet saxifrage *Pimpinella saxifraga*

Butcher's broom *Ruscus aculeatus*

Butterfly bush *Buddleia davidii*

Calamint (see Common Calamint)

Canadian fleabane *Conyza canadensis*

Cat's ear *Hypochoeris radicata*

Charlock *Sinapis arvensis*

Cherry,
    Bird *Prunus padus*
    Wild (Gean) *Prunus avium*

Cherry laurel *Prunus laurocerasus*

Cherry-plum *Prunus cerasifera*

Chickweed (see Common Chickweed)

Cleavers *Galium aparine*

Club-rush *Scirpus lacustris*

Clustered harebell *Campanula glomerata*

Cocksfoot *Dactylis glomerata*

Common calamint *Calamintha sylvatica*

Common chickweed *Stellaria media*

Common horsetail *Equisetum arvense*

Common reed *Phragmites australis*

Common restharrow *Ononis repens*

Common rock-rose *Helianthemum chamaecistus*

Common St John's-wort *Hypericum perforatum*

Common water crowfoot *Ranunculus aquatalis*

Corn buttercup *Ranunculus arvensis*

Corn cockle *Agrostemma githago*

Corn crowfoot (see Corn buttercup)

Cornflower *Centaurea cyanus*

Couch grass (Couch, Twitch) *Elymus repens*

Cowslip *Primula veris*

Crab apple *Malus sylvestris*

Creeping soft-grass *Holcus mollis*

Crested dogstail *Cynosurus cristatus*

Crested hair grass *Koeleria cristata*

Cupressus *Cupressus* spp

Damson (see Plum)

Dog rose (see Rose)

Dog's mercury *Mercurialis perennis*

Dogwood *Cornus sanguinea*

Duckweeds *Lemna* spp

Elder *Sambucus nigra*

Elm,
    Cornish *Ulmus stricta*
    Dutch, Huntingdon *Ulmus* x *hollandica*
    East Anglian *Ulmus minor*
    English *Ulmus procera*
    Wych *Ulmus glabra*

Escallonia *Escallonia macrantha*

European larch *Larix decidua*

Everlasting pea *Lathyrus latifolius*

Fairy flax *Linum catharticum*

False brome *Brachypodium sylvaticum*

False oat-grass *Arrhenatherum elatius*

Fat-hen *Chenopodium album*

Fescues *Festuca* spp

Fescue,

    Red *Festuca rubra*

    Sheep's *Festuca ovina*

    Tall *Festuca arundinacea*

Field horsetail (see Common horsetail)

Field maple *Acer campestre*

Field pansy *Viola arvensis*

Flowering rush *Butomus umbellatus*

Fluellen,

    Round-leaved *Kickxia spuria*

    Sharp-leaved *Kickxia elatine*

Fool's parsley *Aethusa cynapium*

Fool's watercress *Apium nodiflorum*

Forget-me-nots *Myosotis* spp

Frog-bit *Hydrocharis morsus-ranae*

Fuchsia *Fuchsia magellanica*

Fumitory *Fumaria officinalis*

Furze (see Gorse)

Garlic mustard *Alliaria petiolata*

Gean (see Cherry, Wild)

Germander speedwell *Veronica chamaedrys*

Giant water dock (Great water dock) *Rumex hydrolapathum*

Gorse *Ulex europaeus*

Great sallow (Pussy willow, Goat willow) *Salix caprea*

Green helleborine *Epipactis phyllanthes*

Guelder rose *Viburnum opulus*

Harebell *Campanula rotundifolia*

Hart's tongue fern *Asplenium scolopendrium*

Hawthorn,

    Common (Hedgerow) *Crataegus monogyna*

    Midland (Woodland) *Crataegus laevigata*

Hazel *Corylus avellana*

Hedge-bedstraw *Galium mollugo*

Hedge garlic (see Garlic mustard)

Hedge mustard *Sisymbrium officinale*

Hedge woundwort *Stachys sylvatica*

Henbit dead-nettle *Lamium amplexicaule*

Herb Robert *Geranium robertianum*

Hogweed *Heracleum sphondylium*

Holly *Ilex aquifolium*

Honeysuckle *Lonicera periclymenum*

Hop *Humulus lupulus*

Hornbeam *Carpinus betulus*

Horse chestnut *Aesculus hippocastanum*

Horseshoe vetch *Hippocrepis comosa*

Horsetail (see Common horsetail)

Ivy *Hedera helix*

Japanese knotweed *Reynoutria japonica*

Juniper *Juniperus communis*

Kidney vetch *Anthyllis vulneraria*

Kingcup *Caltha palustris*

Knapweed,

    Common *Centaurea nigra*

    Greater *Centaurea scabiosa*

Knapweed broomrape *Orobanche elatior*

Knotgrass *Polygonum aviculare*

Laburnum *Laburnum anagyroides*

Lady's bedstraw *Galium verum*

Lady's smock *Cardamine pratensis*

Larch (see European larch)

Lime,

    Common *Tilia x vulgaris*

    Large-leaved *Tilia platyphyllos*

    Small-leaved (Pry) *Tilia cordata*

Lords and ladies *Arum maculatum*

Lupins *Lupinus* spp

Maple (see Field maple)

Marjoram *Origanum vulgare*

Marsh marigold (see Kingcup)

Mayweed,

    Scented *Matricaria recutita*

    Scentless *Tripleurospermum inodorum*

Meadow-grasses *Poa* spp

Meadow-grass,

    Annual *Poa annua*

    Rough *Poa trivialis*

Meadow vetchling *Lathyrus pratensis*

Mountain ash (see Rowan)

Mouse-ear hawkweed *Hieracium pilosella*

Musk mallow *Malva moschata*

Nettle (see Stinging nettle)

Night-flowering catchfly *Silene noctiflora*

Oak,

    Common (Pedunculate) *Quercus robur*

    Sessile *Quercus petraea*

Old man's beard *Clematis vitalba*

Orache *Atriplex patula*

Ox-eye daisy *Leucanthemum vulgare*

Oxford ragwort *Senecio squalidus*

Parsley-piert *Aphanes arvensis*

Pear *Pyrus communis*

Perennial ryegrass *Lolium perenne*

Pine (see Scots pine)

Pittosporum *Pittosporum crassifolium*

Plum *Prunus domestica*

Poplar,

    Aspen *Populus tremula*

    Black *Populus nigra*

    Lombardy *Populus italica*

    White *Populus alba*

Poppies *Papaver* spp

Poppy,

    Common *Papaver rhoeas*

    Prickly *Papaver argemone*

Primrose *Primula vulgaris*

Privet *Ligustrum vulgare*

Pry (see Small-leaved lime)

Purple-loosestrife *Lythrum salicaria*

Quaking grass *Briza media*

Ratstail fescue *Vulpia myuros*

Red campion *Silene dioica*

Redshank *Polygonum persicaria*

Reed (see Common reed)

Reedmaces *Typha* spp

Reed sweet-grass *Glyceria maxima*

Reflexed salt-marsh-grass *Puccinellia distans*

Restharrow (see Common restharrow)

Rhododendron *Rhododendron ponticum*

River water-dropwort *Oenanthe fluviatilis*

Rock-rose (see Common rock-rose)

Roses *Rosa* spp

Rose,
    Dog *Rosa canina*
    Field *Rosa arvensis*

Rowan (Mountain ash) *Sorbus aucuparia*

Ryegrass,
    Italian *Lolium multiflorum*
    Perennial *Lolium perenne*

Salad burnet *Sanguisorba minor*

Sallow *Salix atrocinerea*

Saxifrage,
    Meadow *Saxifraga granulata*
    Rue-leaved *Saxifraga tridactylites*

Scabious,
    Field *Knautia arvensis*
    Small *Scabiosa columbaria*

Scots pine *Pinus sylvestris*

Sea aster *Aster tripolium*

Sea plantain *Plantago maritima*

Self-heal *Prunella vulgaris*

Service tree *Sorbus torminalis*

Shepherd's-needle *Scandix pecten-veneris*

Sloe (see Blackthorn)

Small toadflax *Chaenorhinum minus*

Sorrel,
    Common *Rumex acetosa*
    Sheep's *Rumex acetosella*

Speedwells *Veronica* spp

Spiked water milfoil *Muriophyllum spicatum*

Spindle *Euonymus europaeus*

Stinging nettle *Urtica dioica*

St John's-wort (see Common St John's-wort)

Stonecrop,
    Biting *Sedum acre*
    White *Sedum album*

Sweet briar *Rosa rubiginosa*

Sweet chestnut *Castanea sativa*

Sweet vernal grass *Anthoxanthum odoratum*

Sycamore *Acer pseudoplatanus*

Thistles *Cirsium* spp

Thorn (see Hawthorn)

Thyme *Thymus praecox*

Timothy *Phleum pratense*

Toadflax,
    Purple *Linaria purpurea*
    Small *Chaenorhinum minus*

Tufted hair-grass *Deschampsia cespitosa*

Unbranched bur-reed *Sparganium emersum*

Venus'-looking-glass *Legousia hybrida*

Water crowfoot (see Common water crowfoot)

Water figwort *Scrophularia auriculata*

Water forget-me-not *Myosotis scorpoides*

Water-lily,
    Fringed *Nymphoides peltata*
    White *Nymphaea alba*
    Yellow *Nuphar lutea*

Water plantain *Alisma plantago-aquatica*

Water violet *Hottonia palustris*

Wayfaring tree *Viburnum lantana*

Weasel's snout (Lesser snapdragon) *Misopates orontium*

Whitebeam *Sorbus aria*

Whitlow grass *Erophila verna*

Wild basil *Clinopodium vulgare*

Wild carrot *Daucus carota*

Wild cherry (see Cherry)

Wild mignonette *Reseda lutea*

Wild oat *Avena fatua*

Wild plum (see Plum)

Wild rose (see Roses)

Willows *Salix* spp

Willow,
    Almond *Salix triandra*
    Bay *Salix pentandra*
    Crack *Salix fragilis*
    Goat *Salix caprea*
    Grey *Salix cinerea*
    Osier *Salix viminalis*
    Purple *Salix purpurea*
    White *Salix alba*

Willowherb,
    American *Epilobium ciliatum*
    Hairy (Hoary) *Epilobium parviflorum*
    Rose-bay *Chamaenerion angustifolium*

Yellow flag *Iris pseudacorus*

Yellow oat-grass *Trisetum flavescens*

Yew *Taxus baccata*

Yorkshire fog *Holcus lanatus*

Animals

Apple twig cutter *Rhynchites caeruleus*

Badger *Meles meles*

Banded agrion *Agrion splendens*

Bank vole *Clethrionomys glareolus*

Barn owl *Tyto alba*

Black bean aphid *Aphis fabae*

Blackbird *Turdus merula*

Blackcap *Sylvia atricapilla*

Black-kneed capsid *Blepharidopterus angulatus*

Blue tit *Parus caeruleus*

Brassica pod midge *Dasineura brassicae*

Brimstone *Gonepteryx rhamni*

Brown hairstreak *Thecla betulae*

Brown hawker *Aeshna grandis*

Bullfinch *Pyrrhula pyrrhula*

Buzzard *Buteo buteo*

Chaffinch *Fringilla coelebs*

Chiffchaff *Phylloscopus collybita*

Cirl bunting *Emberiza cirlus*

Coal tit *Parus ater*

Collared dove *Steptopelia decaocto*

Comma *Polygonia c-album*

Common blue *Polyommatus icarus*

Common shrew (see Shrew)

Common vole *Microtus arvalis*

Coot *Fulica atra*

Corn bunting *Emberiza calandra*

Crow,
    Carrion *Corvus corone corone*
    Hooded *Corvus corone cornix*

Cuckoo *Cuculus canorus*

Curlew *Numenius arquata*

Damson hop aphid *Phorodon humuli*

Dunnock (see Hedge sparrow)

Earwig *Forficula auricularia*

Eel *Anguilla anguilla*

Essex skipper *Thymelicus lineola*

European red mite *Panonychus ulmi*

Fieldfare *Turdus pilaris*

Field mouse (see Woodmouse)

Field vole *Microtus agrestis*

Fox *Vulpes vulpes*

Frog *Rana temporaria*

Garden warbler *Sylvia borin*

Gatekeeper *Pyronia tithonus*

Glow-worm *Lampyris noctiluca*

Goldfinch *Carduelis carduelis*

Grasshopper warbler *Locustella naevia*

Great tit *Parus major*

Greenfinch *Carduelis chloris*

Green-veined white *Pieris napi*

Grey partridge *Perdix perdix*

Hare (Brown) *Lepus capensis*

Hedge brown (see Gatekeeper)

Hedgehog *Erinaceus europaeus*

Hedge sparrow (see Sparrow)

Holly blue *Celastrina argiolus*

House sparrow (see Sparrow)

Jackdaw *Corvus monedula*

Kestrel *Falco tinnunculus*

Kingfisher *Alcedo atthis*

Lapwing *Vanellus vanellus*

Large skipper *Ochlodes venata*

Large white *Pieris brassicae*

Lesser redpoll *Acanthis flammea*

Lesser whitethroat *Sylvia curruca*

Lettuce root aphid *Phemphigus bursarius*

Linnet *Acanthis cannabina*

Little owl *Athene noctua*

Long-tailed tit *Aegithalos caudatus*

Magpie *Pica pica*

Marsh tit *Parus palustris*

Meadow brown *Maniola jurtina*

Meadow pipit *Anthus pratensis*

Mistle thrush (see Thrush)

Mole *Talpa europaea*

Moorhen *Gallinula chloropus*

Nightingale *Luscinia megarhynchos*

Nuthatch *Sitta europaea*

Orange tip *Anthocharis cardamines*

Painted lady *Vanessa cardui*

Partridge (see Grey partridge)

Pea and bean weevil *Sitonia lineatus*

Peach potato aphid *Myzus persicae*

Peacock butterfly *Inachis io*

Pheasant *Phasianus colchicus*

Pied flycatcher *Ficedula hypoleuca*

Pygmy shrew (see Shrew)

Rabbit *Oryctolagus cuniculus*

Rat (Common or Brown) *Rattus norvegicus*

Red admiral *Vanessa atalanta*

Red fox (see Fox)

Red-backed shrike *Lanius cristatus*

Red-legged partridge *Alectoris rufa*

Redstart *Phoenicurus phoenicurus*

Redwing *Turdus iliacus*

Reed bunting *Emberiza schoeniclus*

Ringlet *Aphantopus hyperantus*

Roach *Rutilus rutilus*

Robin *Erithacus rubecula*

Rook *Corvus frugilegus*

Sand martin *Riparia riparia*

Sedge warbler *Acrocephalus schoenobaenus*

Seven spot ladybird *Coccinella 7-punctata*

Shrew,
> Common *Sorex araneus*
> Pygmy *Sorex minutus*

Skylark *Alauda arvensis*

Small copper *Lycaena phlaeus*

Small heath *Coenonympha pamphilus*

Small skipper *Thymelicus sylvestris*

Small tortoiseshell *Aglais urticae*

Small white *Pieris rapae*

Song thrush (see Thrush)

Sparrow,
> Hedge *Prunella modularis*
> House *Passer domesticus*
> Tree *Passer montanus*

Speckled wood *Pararge aegeria*

Starling *Sturnus vulgaris*

Stoat *Mustela erminea*

Stock dove *Columba oenas*

Stonechat *Saxicola torquata*

Stone curlew *Burhinus oedicnemus*

Tawny owl *Strix aluco*

Thrush,
> Mistle *Turdus viscivorus*
> Song *Turdus philomelos*

Toad *Bufo bufo*

Treecreeper *Certhia familiaris*

Tree sparrow (see Sparrow)

Turtle dove *Streptopelia turtur*

Wall brown *Pararge megaera*

Water vole *Arvicola terrestris*

Weasel *Mustela nivalis*

Whitethroat *Sylvia communis*

Willow tit *Parus montanus*

Willow warbler *Phylloscopus trochilus*

Woodmouse (Long-tailed field mouse) *Apodemus sylvaticus*

Woodpecker,
> Great spotted *Dendrocopos major*
> Green *Picus viridis*
> Lesser spotted *Dendrocopos minor*

Wood pigeon *Columba palumbus*

Wren *Troglodytes troglodytes*

Yellowhammer *Emberiza citrinella*

ACKNOWLEDGEMENTS

The Open University course team is greatly indebted to the many people, with a wide range of experience of countryside management, who have contributed to the development of this teaching programme.

We acknowledge the very generous financial support of the former Nature Conservancy Council, along with the Esmée Fairbairn Charitable Trust and the Ernest Cook Trust.

We thank our external assessor, Professor Bryn Green, The Sir Cyril Kleinwort Professor of Countryside Management, Wye College, University of London, for his valuable comments and support.

We are extremely grateful to R. Deane (Senior Adviser, Farming and Wildlife Advisory Group) who provided the material for the case study farm, and to A. Derby, the owner of the farm, and P. Doble, the farm manager, for their interest and involvement.

We also thank the many other people who provided source material for the book and read and commented on preliminary drafts:

S. Alston (Farming and Wildlife Advisory Group)

J. Comont (Bedfordshire County Council)

H. Currie (Scottish Conservation Projects Trust)

R. Grant (Countryside Commission for Scotland)

J. Greenall (Farming and Wildlife Advisory Group)

P. Hallam (Dry Stone Walling Association)

T. Hammond (English Nature)

J. Harvey (The National Trust)

J. Hughes (Farming and Wildlife Advisory Group)

K. Jones (Agricultural Development and Advisory Service)

G. Kerby (Agricultural Training Board)

R. Knight (Farming and Wildlife Advisory Group)

P. Knipe (Hampshire Lineside Vegetation Project)

R. MacMullen (Farming and Wildlife Advisory Group)

J. Marshall (Long Ashton Research Station)

N. Moore (formerly Nature Conservancy Council)

A. Moorhouse (Farming and Wildlife Advisory Group)

R. Robinson (Nature Conservancy Council, Scotland)

J. Simpkins (Dry Stone Walling Association)

H. Smith (University of Oxford)

N. Sotherton (The Game Conservancy)

P. Spencer (Farming and Wildlife Advisory Group)

M. Thomas (University of Southampton)

Grateful acknowledgement is made to the following sources for permission to use material in this book:

Figures and boxes

Figure 1.4 (top left): Rural Resource Management Department, The Scottish Agricultural College, Edinburgh; *Figure 1.4 (bottom left)*: Steve Hartgroves, Cornwall Archaeological Unit; *Figure 1.6*: Gilbert, O. (1989) *The Ecology of Urban Habitats*, Chapman & Hall, © Oliver Gilbert, 1989; *Figure 1.7 (except top right)*: Cambridge University Collection; *Figure 1.7 (top right)*: Robert White, Yorkshire Dales National Park; *Figure 2.1*: Cobham Resource Consultants (1985) *Landscape and Wildlife Conservation on Farms*, Countryside Commission for Scotland; *Figure 2.2 (top right)*: Pat Harvey, Manor Farm, Ourston, Leicestershire; *Figure 2.4 (photograph)*: Robert Deane, Hereford and Worcester Farming and Wildlife Advisory Group; *Boxes 2.1, 4.1, 5.1, 6.2*: Adapted by kind permission of the British Trust for Conservation Volunteers; *Box 4.1 and Figure 4.1*: Adapted from Sturrock, F. G. and Cathie, J. (1980) *Farm Modernisation and the Countryside*, Department of Land Economy, University of Cambridge; *Figure 5.3*: Adapted by kind permission of the British Society for Conservation Volunteers; *Figure 5.5*: Reproduced by courtesy of the Forestry Commission; *Figure 6.4*: Harris, J., Jefferies, D. and Creswell, W. (1990) *Problems With Badgers?*, Royal Society for the Prevention of Cruelty to Animals.

INDEX

Scottish Conservation Projects Trust 68, 74, 110, 111

Severn Trent Water Authority 103

shade 34, 53

shelter-belts 9, 118: hedges 5, 81; trees 86, 87

shrubs 9, 81: hedgerow 33; wildlife habitats 35, 40

Sites of Special Scientific Interest (SSSI) 44, 118

soil conditions, effect on species distribution and diversity 33, 90, 91

soil erosion reduction 50, 53

spraying 55, 61, 107, 108: safety precautions 111; *see also* fertilisers, fungicides, herbicides, pesticides

stock, see livestock

stockproof barriers 10: ditches 57; hedges 68, 81, 83

stone walling 54, 74, 92, 93: labour requirements 106; tools 107; training 74, 111

stone walls 5, 33; building, maintenance, *see* stone walling; colour 27; case study farm 48, 103; landscape value 5, 23, 24, 27; materials 74, 93; wildlife habitats 34, 48

streams 48, 62; *see also* water courses

strip cultivation 14, 118

Suffolk, Bronze Age field boundaries 12

Sussex, lynchets 12

T

timber and wood products 58, 60: from coppicing 71; from hedgerow trees 73, 86

tools 107–9: safety precautions 111; training 68, 74, 111

tourism 60

towpaths 11, 12, 18, 118

tracks 47, 101, 102

training courses 68, 74, 111

transport routes 5, 60, 94: landscape value 28, 29; *see also* canals, motorways, railways, roads

Tree Preservation Order (TPO) 28, 73, 118

trees: banks 57; hedgerow, *see* hedgerow trees; felling 73, 79; management and maintenance 72, 73, 79, 85; new planting 72, 73, 84, 94; timber and wood products 58, 71, 73; transport verges 57, 79, 94

trimming, hedge 65, 66: frequency and timing 53, 65, 66, 99, 106; labour requirements 53, 106, 111; new planting 83, 84; tools 106, 107

turnpike roads 18, 118

V

verges 6, 11; management and maintenance 11, 57, 58, 78, 79, 112

vermin 57

W

Wales: ancient field boundaries 10, 15; hedge laying styles 69; hedge shapes 67

walkers: case study farm 19, 30, 61, 102; *see also* footpaths, recreational land use

walls, see stone walls

water courses 75, 76: pollution 34, 112; *see also* canals, ditches, rivers, streams

weeds 47, 55: control 75, 77, 84, 88, 90, 91 (herbicide use) 72, 84, 87; case study farm 61, 101, 102; *see also* arable weed conservation, arable weed margins

wild flowers 79, 84, 87; management and maintenance 78, 79, 95; perennial grass 88, 90 (case study farm) 47, 101, 102

wildlife habitats 33, 34, 38; *see also individual boundary types, e.g.* hedges

wildlife trusts 44, 110

Wiltshire: ancient field boundaries 12; hedge laying styles 70; lynchets 12; planned countryside 15

windbreaks, *see* shelter-belts

wood products, *see* timber and wood products

Y

Yorkshire: field boundaries 13; inset field barns 5; lynchets 12; planned countryside 15